Current Topics
in Membranes and Transport

VOLUME 23

Genes and Membranes: Transport Proteins and Receptors

Contributors

Giovanna Ferro-Luzzi Ames
Sandra Barbel
Sylvie Bedu
Alan Boyd
William A. Catterall
Maria Costa
Carolyn Doyle
Wolfgang Epstein
Mary-Jane Gething
Tohru Gonoi
Lily Yeh Jan
Yuh Nung Jan
Alexandra Krikos
Ching Kung
Cheryl Laffer
Robert Levenson
Benno Müller-Hill
Norihiro Mutoh
Patrick O'Farrell
Resha M. Putzrath
Michael Roth
Yoshiro Saimi
Lawrence Salkoff
Joe Sambrook
A. E. Senior
Donna Seto-Young
Melvin I. Simon
Leslie Timpe
T. Hastings Wilson

Current Topics in Membranes and Transport

Edited by

Felix Bronner
Department of Oral Biology
University of Connecticut Health Center
Farmington, Connecticut

Arnost Kleinzeller
Department of Physiology
University of Pennsylvania School of Medicine
Philadelphia, Pennsylvania

VOLUME 23

Genes and Membranes: Transport Proteins and Receptors

Guest Editors

Edward A. Adelberg
Department of Human Genetics
Yale University School of Medicine
New Haven, Connecticut

Carolyn W. Slayman
Departments of Human Genetics and Physiology
Yale University School of Medicine
New Haven, Connecticut

Volume 23 is part of the series (p. xiii) from the Yale Department of Physiology under the editorial supervision of:

Joseph F. Hoffman
Department of Physiology
Yale University School of Medicine
New Haven, Connecticut

Gerhard Giebisch
Department of Physiology
Yale University School of Medicine
New Haven, Connecticut

1985

ACADEMIC PRESS, INC.
(Harcourt Brace Jovanovich, Publishers)
Orlando San Diego New York London
Toronto Montreal Sydney Tokyo

ACADEMIC PRESS, INC.
Orlando, Florida 32887

United Kingdom Edition published by
ACADEMIC PRESS INC. (LONDON) LTD.
24–28 Oval Road, London NW1 7DX

LIBRARY OF CONGRESS CATALOG CARD NUMBER: 70-117091

ISBN 0-12-153323-9

PRINTED IN THE UNITED STATES OF AMERICA

85 86 87 88 9 8 7 6 5 4 3 2 1

Contents

PART II. CHANNELS

CHAPTER 3. Ca^{2+} Channels of *Paramecium:* A Multidisciplinary Study

CHING KUNG AND YOSHIRO SAIMI

CHAPTER 4. Studies of *Shaker* Mutations Affecting a K^+ Channel in *Drosophila*

LILY YEH JAN, SANDRA BARBEL, LESLIE TIMPE, CHERYL LAFFER, LAWRENCE SALKOFF, PATRICK O'FARRELL, AND YUH NUNG JAN

CHAPTER 5. Sodium Channels in Neural Cells: Molecular Properties and Analysis of Mutants

WILLIAM A. CATTERALL, TOHRU GONOI, AND MARIA COSTA

PART III. TRANSPORT SYSTEMS

CHAPTER 6. The Histidine Transport System of *Salmonella typhimurium*

GIOVANNA FERRO-LUZZI AMES

CHAPTER 7. A Study of Mutants of the Lactose Transport System of *Escherichia coli*

T. HASTINGS WILSON, DONNA SETO-YOUNG, SYLVIE BEDU, RESHA M. PUTZRATH, AND BENNO MÜLLER-HILL

CHAPTER 8. The Proton-ATPase of *Escherichia coli*

A. E. SENIOR

CHAPTER 9. **The Kdp System: A Bacterial K^+ Transport ATPase**

WOLFGANG EPSTEIN

CHAPTER 10. **Molecular Cloning and Characterization of a Mouse Ouabain Resistance Gene: A Genetic Approach to the Analysis of the Na^+,K^+-ATPase**

ROBERT LEVENSON

Contributors

Numbers in parentheses indicate the pages on which the authors' contributions begin.

Giovanna Ferro-Luzzi Ames, Department of Biochemistry, University of California, Berkeley, California 94720 (103)

Sandra Barbel, Department of Physiology, University of California, San Francisco, California 94143 (67)

Sylvie Bedu,[1] Department of Physiology, Harvard Medical School, Boston, Massachusetts 02115 (121)

Alan Boyd, Leicester Biocentre, Medical Sciences Building, Leicester, England (3)

William A. Catterall, Department of Pharmacology, University of Washington, Seattle, Washington 98195 (79)

Maria Costa,[2] Department of Pharmacology, University of Washington, Seattle, Washington 98195 (79)

Carolyn Doyle, Cold Spring Harbor Laboratory, Cold Spring Harbor, New York 11724 (17)

Wolfgang Epstein, Department of Molecular Genetics and Cell Biology, University of Chicago, Chicago, Illinois 60637 (153)

Mary-Jane Gething, Cold Spring Harbor Laboratory, Cold Spring Harbor, New York 11724 (17)

Tohru Gonoi, Department of Pharmacology, University of Washington, Seattle, Washington 98195 (79)

Lily Yeh Jan, Department of Physiology, University of California, San Francisco, California 94143 (67)

Yuh Nung Jan, Department of Physiology, University of California, San Francisco, California 94143 (67)

Alexandra Krikos, Division of Biology, California Institute of Technology, Pasadena, California 91125 (3)

Ching Kung, Laboratory of Molecular Biology and Department of Genetics, University of Wisconsin, Madison, Wisconsin 53706 (45)

Cheryl Laffer, Department of Physiology, University of California, San Francisco, California 94143 (67)

Robert Levenson,[3] Center for Cancer Research and Department of Biology, Massachusetts Institute of Technology, Cambridge, Massachusetts 02139 (177)

Benno Müller-Hill, Institut für Genetik der Universität zu Köln, 5000 Cologne 41, Federal Republic of Germany (121)

Norihiro Mutoh, Division of Biology, California Institute of Technology, Pasadena, California 91125 (3)

[1]Present address: U. E. R. Scientifique Marseille-Luminy, Physiologie Cellulaire, 13288 Marseille Cedex 9, France.

[2]Present address: Department of Hematology, University of Rochester, School of Medicine and Dentistry, Rochester, New York 14642.

[3]Present address: Department of Cell Biology, Yale University School of Medicine, New Haven, Connecticut 06510.

Patrick O'Farrell, Department of Biochemistry, University of California, San Francisco, California 94143 (67)

Resha M. Putzrath, Environ Corporation, 777 Fourteenth Street N.W., Suite 1000, Washington, D.C. 20005 (121)

Michael Roth, Cold Spring Harbor Laboratory, Cold Spring Harbor, New York 11724 (17)

Yoshiro Saimi, Laboratory of Molecular Biology, University of Wisconsin, Madison, Wisconsin 53706 (45)

Lawrence Salkoff, Department of Anatomy and Neurobiology, Washington University School of Medicine, St. Louis, Missouri 63110 (67)

Joe Sambrook, Cold Spring Harbor Laboratory, Cold Spring Harbor, New York 11724 (17)

A. E. Senior, Department of Biochemistry, University of Rochester Medical Center, Rochester, New York 14642 (135)

Donna Seto-Young,[4] Department of Physiology, Harvard Medical School, Boston, Massachusetts 02115 (121)

Melvin I. Simon, Division of Biology, California Institute of Technology, Pasadena, California 91125 (3)

Leslie Timpe, Department of Physiology, University of California, San Francisco, California 94143 (67)

T. Hastings Wilson, Department of Physiology, Harvard Medical School, Boston, Massachusetts 02115 (121)

[4]Present address: National Research Council of Canada, Division of Biological Sciences, Molecular Genetics Section, Ottawa, Ontario K1A OR6, Canada.

Preface

A major aim of this volume is to collate in one source the work of investigators working on membrane proteins in a wide variety of cells and tissues, who share an interest in the use of new genetic technologies to probe structure–function relationships. Each of the three parts of this volume reflects this aim. Part I covers genetic studies on receptors and recognition proteins, both in bacteria and in animal viruses. Part II treats the genetic control of channel proteins in *Paramecium, Drosophila,* and cultured neural cells. In Part III, the emphasis shifts to the genetics of bacterial and mammalian transport systems.

We are grateful to the members of the Yale University Department of Physiology who worked to make possible the Ninth Conference on Membrane Transport Processes, which provided the basis for this volume. We express special gratitude to Rita Scott and Marie Santore who helped with every phase of the arrangements. Generous financial support was provided by Abbott Laboratories, Bayer AG/Miles, Biogen Research Corporation, E. I. du Pont de Nemours and Co., Inc., and The Revlon Health Care Group.

EDWARD A. ADELBERG
CAROLYN W. SLAYMAN

Yale Membrane Transport Processes Volumes

Joseph F. Hoffman (ed). (1978). "Membrane Transport Processes," Vol. 1. Raven, New York.

Daniel C. Tosteson, Yu. A. Ovchinnikov, and Ramon Latorre (eds.). (1978). "Membrane Transport Processes," Vol. 2. Raven, New York.

Charles F. Stevens and Richard W. Tsien (eds.). (1979). "Membrane Transport Processes," Vol. 3: Ion Permeation through Membrane Channels. Raven, New York.

Emile L. Boulpaep (ed.). (1980). "Cellular Mechanisms of Renal Tubular Ion Transport": Volume 13 of *Current Topics in Membranes and Transport* (F. Bronner and A. Kleinzeller, eds.). Academic Press, New York.

William H. Miller (ed.). (1981). "Molecular Mechanisms of Photoreceptor Transduction": Volume 15 of *Current Topics in Membranes and Transport* (F. Bronner and A. Kleinzeller, eds.). Academic Press, New York.

Clifford L. Slayman (ed.). (1982). "Electrogenic Ion Pumps": Volume 16 of *Current Topics in Membranes and Transport* (A. Kleinzeller and F. Bronner, eds.). Academic Press, New York.

Joseph F. Hoffman and Bliss Forbush III (eds.). (1983). "Structure, Mechanism, and Function of the Na/K Pump": Volume 19 of *Current Topics in Membranes and Transport* (F. Bronner and A. Kleinzeller, eds.). Academic Press, New York.

James B. Wade and Simon A. Lewis (eds.). (1984). "Molecular Approaches to Epithelial Transport": Volume 20 of *Current Topics in Membranes and Transport* (A. Kleinzeller and F. Bronner, eds.). Academic Press, New York.

Edward A. Adelberg and Carolyn W. Slayman (eds.). (1985). "Genes and Membranes: Transport Proteins and Receptors": Volume 23 of *Current Topics in Membranes and Transport* (F. Bronner and A. Kleinzeller, eds.). Academic Press, New York.

Part I

Receptors and Recognition Proteins

Chapter 1

Sensory Transduction in Bacteria

MELVIN I. SIMON, ALEXANDRA KRIKOS,**
NORIHIRO MUTOH, AND ALAN BOYD†*

**Division of Biology*
California Institute of Technology
Pasadena, California

and

†Leicester Biocentre
Leicester, England

I. INTRODUCTION

A variety of bacteria are motile, swimming through the medium by virtue of flagella, which are long (10–12 μm), rigid, helical filaments driven to rotate by a complex structure at the base of the flagellar organelle. A peritricious cell may have 8 to 10 flagella. These can form a "bundle" at one end of the cell, and they then rotate in a coordinated way and thus propel the bacterium through the medium. When the flagella rotate in one direction, e.g., counterclockwise, the organism swims smoothly. However, when some of the flagella change direction of rotation (e.g., switch momentarily to clockwise rotation), the flagella bundle may "jam" and the cell will tumble in a disoriented fashion. When counterclockwise rotation resumes, the cell can reorient its direction of swimming.

ISBN 0-12-153323-9

Therefore, when bacterial behavior is observed in the microscope, cells are seen to undergo a series of "runs" and "tumbles" (for reviews, see Springer *et al.*, 1979; Koshland, 1979; Berg, 1975; Silverman and Simon, 1977b; Macnab, 1978). In an isotropic chemical environment, the cells show random runs and tumbles. However, in the presence of a concentration gradient of a chemical that acts as an attractant, the swimming behavior is found to be biased. Bacteria swimming in the direction of increasing attractant concentration suppress tumbles and thus continue to swim in that direction, whereas bacteria swimming in the opposite direction, i.e., toward lower concentrations of attractant, show an increased frequency of tumbles. The net result of this bias is that the cells migrate in the direction of increasing attractant concentration. Thus, by regulating the frequency of flagellar rotation reversal, the cell can control motility so that it swims in a responsive way. Responsiveness can be studied by tethering the bacterial flagellum to a glass slide. When the flagellar filament is rigidly bound to the surface, the bacterium is driven to rotate (Berg and Anderson, 1973; Silverman and Simon, 1974; Larsen *et al.*, 1974). By observing cellular rotation in response to the addition or removal of attractants from the medium, a great deal can be learned about the physiology of the chemosensory response. These kinds of studies coupled with classical bacterial genetics have resulted in the emergence of a clear picture of the components of the chemosensory pathway.

One of the first steps in the chemosensory pathway is the detection of changes in concentrations of specific attractant and repellent compounds (Adler, 1969). This process is mediated by ligand-binding proteins that are found in the periplasmic space or as transmembrane proteins spanning the cytoplasmic membrane (Silverman and Simon, 1977a; Springer *et al.*, 1977). They act as receptors and directly bind specific sugars or amino acids. The periplasmic ligand-binding proteins may further interact with transmembrane proteins. The proteins that span the cytoplasmic membrane initiate a change that is eventually transduced into a signal that affects the frequency and direction of flagellar rotation.

A number of the transducing proteins have been studied in detail. These have been shown to be multifunctional. They can act as receptors and bind specific ligands and they also function as signal transducing proteins. Thus far, in *Escherichia coli,* four such transducer genes have been identified. They are *tar, tsr, tap,* and *trg* (for review, see Boyd and Simon, 1982). These genes all encode transmembrane proteins and each of them appears to be responsible for the transduction of signals that result from the binding of a specific subset of attractants and repellents. Some of these proteins have themselves been shown to be receptors and bind specific attractants (Clarke and Koshland, 1979; Hedblom and Adler, 1980). Thus, the Tsr protein is both the receptor for the attractant serine and the transducer which processes signals resulting from changes in the external concentration of serine in the environment. The Tar protein is the receptor for aspartate. It binds aspartate and is also the transducing protein that generates signals in response to changes in the external concentration of aspartate in the environment. The *trg* gene specifies a protein that acts as a transducer for changes in the concentrations of the sugars galactose and ribose (Hazelbauer and Parkinson, 1977). However, the *trg* gene product does not itself bind these sugars. Rather, specific sugar-binding proteins found in the periplasm bind the sugars, and these polypeptides then interact with the *trg* gene product. The *trg* gene then functions to transduce signals resulting from changes in concentration of sugars in the environment. The precise specificity and role of the *tap* gene product are not known. However, it has all of the salient features of the other transducer genes (Wang and Koshland, 1980; Boyd *et al.*, 1981).

In addition to acting as receptors and initiating the signal that regulates flagellar rotation, the transducer proteins also play a central role in information processing, i.e., modulating the effects of ligand binding so that the organism responds to change in attractant concentration as a function of time (Macnab and Koshland, 1972). It is clear that bacteria are able to adapt to changes in attractant concentration. Thus, in the presence of high ambient levels of attractant or very low levels of attractant, the cells are still able to show responsiveness. If attractant concentration is raised rapidly, the cells respond by suppressing flagellar reversal. However, after a few minutes they begin to adapt to the higher concentration of attractant and flagellar reversal begins to occur. The adaptation process has been shown to correlate with the ability of the transducer molecule to be methylated at specific glutamic acid residues in the cytoplasmic portion of the transducer protein (Springer *et al.*, 1979). In summary, therefore, these receptor–transducer proteins may initiate and participate in at least three processes that are central to chemotaxis: (1) they act as receptors and are able to bind ligands on the outer surface of the cytoplasmic membrane or they interact with specific periplasmic ligand-binding proteins; (2) they are responsible for the transmembrane transduction of the ligand-binding events and the initiation of an excitatory signal that affects flagellar rotation; and (3) they participate in the adaptation process and are themselves transiently modified by specific methylation of glutamyl residues on the inner cytoplasmic portion of the transducer polypeptide.

The observations related to transducer function are all assembled in the model shown in Fig. 1. The transducer proteins Tsr, Tar, Tap, and Trg are transmembrane proteins and have three major functions. One is chemoreceptor and this involves binding specific ligands at the outer surface of the cytoplasmic membrane. These binding events then lead to two processes, one an adaptation process which is accompanied by the methylation of the transducer. Methylation is mediated by the gene products of the *che*R gene, which encodes a methyl transferase, and the *che*B gene, which encodes an enzyme that is a methyl esterase. These lead to a steady-state level of methylation of the transducer. In addition, a signal is generated which then affects the frequency of reversal of

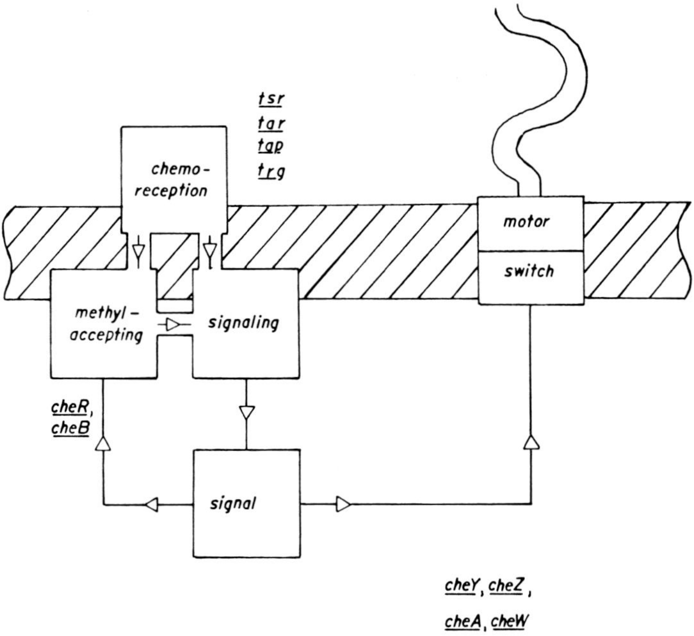

FIG. 1. A model for the function of the receptor–transducer molecules. The cross-hatched region represents the cytoplasmic membranes. The basal region of the flagellar apparatus is represented by two functional regions, the switch at which flagellar reversals are initiated and the motor which drives the flagellar filament to rotate. The receptor–transducers are shown as transmembrane entities with distinct domains embodying each of the molecule's functions.

flagellar rotation. The signaling process appears to be mediated by the products of four genes: *che*Y, *che*Z, *che*A, and *che*W (for review, see Parkinson, 1981; Koshland, 1981). Signaling, methylation, and demethylation of all the transducer proteins are mediated by the same gene products.

In its simplest form, the model for a transducer–receptor shown in Fig. 1 translates into a polypeptide that has three domains. Each domain of the protein would be responsible for one of these basic functions. We would expect that the structural domain involved in adaptation in all of the transducers and the structural domain involved in signaling might have conserved amino acid sequence, since these domains all interact with the same gene products. On the other hand, the receptor domains, i.e., the extracytoplasmic portion of the proteins, might be expected to be divergent since these domains must accommodate the binding of molecules that are sterically very different. Indeed, specific nucleic acid hybridization experiments (Boyd *et al.*, 1981), as well as studies of the interaction of the transducers with specific antibodies (Wang and Koshland, 1980) and comparisons of the products of partial proteolysis of transducer proteins

(Hazelbauer and Engstrom, 1981), all indicate that they are composed of a region of highly conserved amino acid sequence, as well as a region of highly divergent amino acid sequence (Boyd *et al.*, 1981). To probe further the structure and function of these molecules, we isolated the nucleic acid segments that correspond to the *tar, tap,* and *tsr* genes and compared their nucleic acid sequences and thus their amino acid sequences.

II. MAPPING THE TRANSDUCER GENES

In order to define the stretch of DNA that corresponded exactly to each of the transducer genes, transposons were introduced into a target region and their effect on transducer gene function was measured. A plasmid that carried the region including the transducer genes was introduced into a strain that was deficient for all the transducers. The plasmid conferred chemotaxis on the strain. This is seen clearly as a ring of bacteria that migrates from the point of inoculation on a motility plate (Fig. 2). When Tn5 transposons that interrupted the

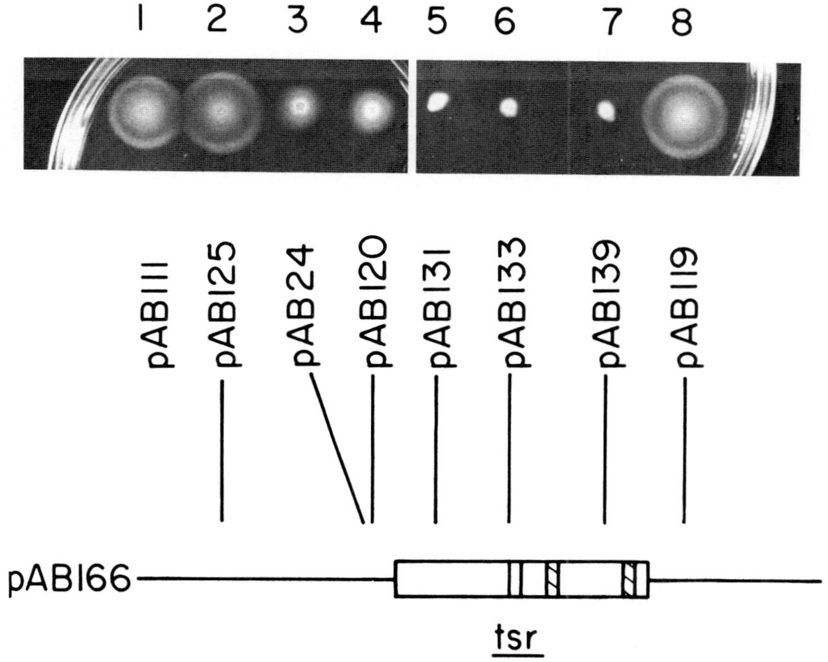

FIG. 2. Mapping the extent of the transducer genes. Transposons are inserted into the plasmid pAB100. The line extending to the map of the plasmid represents the relative position of the inserted transposon as determined by restriction endonuclease mapping. The upper portion of the figure shows the pattern of motility found when strains deficient in *tar* and *tsr* carry the plasmid with transposons inserted near or in the *tsr* gene.

integrity of the transducer gene were introduced, they eliminated the ability of the plasmid to allow chemotaxis. Figure 2 shows the relative positions of some of the transposons as determined by mapping with restriction endonucleases. The ability of the plasmid to confer chemotaxis is shown by its function in the motility agar test assay. The relative effectiveness of the plasmid could be estimated by measuring the migration of the ring of bacteria on the motility plate. Transposons that inserted within the coding region of the *tsr* gene disrupted activity completely and eliminated chemotaxis, whereas those transposons that inserted adjacent to the gene in regulatory regions decreased the chemotaxis activity of the resulting strain. Finally, transposons that inserted in regions that did not encode parts of the gene had no effect on chemotaxis. Thus, the nucleic acid segment responsible for transducer function was clearly defined.

III. THE AMINO ACID SEQUENCES OF THREE TRANSDUCERS

A. Homologous Sequences

The nucleic acid sequence of the *tsr, tar,* and *tap* genes was determined (Boyd *et al.*, 1983; Krikos *et al.*, 1983). Figure 3 compares the amino acid sequence derived from the nucleic acid sequences of the three genes. It is clear that there are a number of regions of highly conserved amino acid sequence. The degree of homology between the amino acid sequences of the *tsr* and *tar* gene products is compared in a graphic way in Fig. 4. There is approximately 60% amino acid identity between the *tar* and *tsr* genes in the sequence from residue 1 to residue 31. Then there are three blocks of sequence that show more than 90% amino acid identity. These include the region from residues 291–313, which specifies an amino acid sequence that includes the glutamate residues that are methylated during chemotaxis. This stretch has been shown (Khery and Dahlquist,1982a,b) to include at least two of the methylation sites. The other methylation sites are found in a conserved region which extends from residues 480 to 502. Table I summarizes the amino acid sequences that surround the methylation site. It is clear that each of the sites has a ''consensus'' sequence. The first two amino acids are either glutamine or glutamic acid, and the second glutamyl residue is the one that is methylated. This is followed by another amino acid which varies and then there is either an alanine or threonine, or an alanine or serine. It is not surprising that glutamines act as methylation sites since it has been shown that the demethylase enzyme (the product of the *che*B gene) is also capable of deamidating specific glutamine residues. Deamidation is required before complete methylation of any of the transducers can occur. It is therefore these glutamines that are first deamidated and then become available for methylation

```
tsr  M L K R I K I V T S L L L V L A V F G L L Q L T S G G L F F N A L K N D K E N F (40)
tar  M I N R I R V V T L L V M V L G V F A L L Q L I S G S L F F S S L H H S Q K S F (40)
tap  M F N R I R I S T T L F L I L I L C G I L Q I G S N G M S F W A F R D D L Q R L (40)

tsr  T V L Q T I R Q Q Q S T L N G S W V A L L Q T R N T L N R A G I R Y M M D Q N N (80)
tar  V V S N Q L R E Q Q G E L T S T W D L M L Q T R I N L S R S A V R M M M D S S N (80)
tap  N Q V E Q S N Q Q R A A L A Q T R A V M L Q A S T A L N K A G T L T A L S Y P A (80)

tsr  I G S G S T V A E L M E S A S I S L K Q A E K N W A D Y E A L P R D P R Q S T A (120)
tar  Q Q S N A K V   E L L D S A R K T L A Q A A T H Y K K F K S M A P L P   E M V A (118)
tap  D D I K T   _   _ L M T T A R A S L T Q S T T L F K S F M A M T A G N E H V R G (116)

tsr  A A A E I K R N Y D I Y H N A L A E L I Q L L G A G K S T S S L I S R P R D I R (160)
tar  T S R N I D E K Y K N Y Y T A L T E L I D Y L D Y G N T G A Y F A Q P T Q G M Q (158)
tap  L Q K E T E K S F A R W H N D L E H Q A T W L E S N Q L S D F L T A P V Q G S Q (156)

tsr  N G F E K Q Y V A Y M E Q N D R L H D I A V S D N N A S Y S Q A M W I L V G V M (200)
tar  N A M G E R F A Q Y A L S S E K L Y R D I V T D N A D D Y R F A Q W Q L A V I A (198)
tap  N A F D V N F E A W Q L E I N H V L E A A S A Q S Q R N Y Q I S A L V F I S M I (196)

tsr  I V V L A V I F A V W F G I K A S L V A P M N R L I D S I R H I A G G D L V K P (240)
tar  L V V V L I L L V A W Y G I R R M L L T P L A K I I A H I R E I A G G N L A N T (238)
tap  I V A A I Y I S S A L W W T R K M I V Q P L A I I G S H F D S I A A G N L A R P (236)

tsr  I E V D G S N E M G Q L A E S L R H M Q G E L M R T V G D V R N G A N A I Y S G (280)
tar  L T I D G R S E M G D L A Q S V S H M Q R S L T D T V T H V R E G S D A I Y A G (278)
tap  I A V Y G R N E I T A I F A S L K T M Q Q A L R G T V S D V R K G S Q E M H I G (276)

tsr  A S E I A T G N N D L S S R T E Q Q A A S L E E T A A S M E Q L T A T V K Q N A (320)
tar  T R E I A A G N T D L S S R T E Q Q A S A L E E T A A S M E Q L T A T V K Q N A (318)
tap  I A E I V A G N N D L S S R T E Q Q A A S L A Q T A A S M E Q L T A T V G Q N A (316)

tsr  E N A R Q A S H L A L S A S E T A Q R G G K V V D N V V Q T M R D I S T S S Q K (360)
tar  D N A R Q A S Q L A Q S A S D T A Q H G G K V V D G V V K T M H E I A D S S K K (358)
tap  D N A R Q A S E L A K N A A T T A Q G G G V Q V S T M T H T M Q E I A T S S Q K (356)

tsr  I A D I I S V I D G I A F Q T N I L A L N A A V E A A R A G E Q G R G F A V V A (400)
tar  I A D I I S V I D G I A F Q T N I L A L N A A V E A A R A G E Q G R G F A V V A (398)
tap  I G D I I S V I D G I A F Q T N I L A L N A A V E A A R A G E Q G R G F A V V A (396)

tsr  G E V R N L A Q R S A Q A A R E I K S L I E D S V G K V D V G S T L V E S A G E (440)
tar  G E V R N L A S R S A Q A A K E I K A L I E D S V S R V D T G S V L V E S A G E (438)
tap  G E V R N L A S R S A Q A A K E I K G L I E E S V N R V Q Q G S K L V N N A A A (436)

tsr  T M A E I V S A V T R V T D I M G E I A S A S D E Q S R G I D Q V G L A V A E M (480)
tar  T M N N I V N A V T R V T D I M G E I A S A S D E Q S R G I D Q V A L A V S E M (478)
tap  T M I D I V S S V T R V N D I M G E I A S A S E E Q Q R G I E Q V A Q A V S Q M (476)

tsr  D R V T Q Q N A A L V E E S A A A A A A L E E Q A S R L T E A V A V F R I Q Q Q (520)
tar  D R V T Q Q N A S L V Q E S A A A A A A L E E Q A S R L T Q A V S A F R L A A S (518)
tap  D Q V T Q Q N A S L V E E A A V A T E Q L A N Q A D R L S S R V A V F T L E E H (516)

tsr  Q R E T S A V V K T V T P A A P                                               (536)
tar  P L T N K P Q T P S R P A S E Q P P A Q P R L R I A E Q D P N W E T F         (553)
tap  E V A R H E S V Q L T N C A S G I L K                                         (535)
```

FIG. 3. The amino acid sequences of the *tsr, tar,* and *tap* gene products. Boxed residues denote positions where the same amino acid is found in all three proteins. Underlining indicates a position at which two of the three amino acid are identical.

TABLE I

METHYLATED SEQUENCES IN THE Tar, Tsr, AND Tap PROTEINS[a]

K1	Tsr 295–317	T E Q Q A A S L E E T A A S M E Q L T A T V K
	Tar 293–315	T E Q Q A S A L E E T A A S M E Q L T A T V K
	Tap 291–313	T E Q Q A A S L A Q T A A S M E Q L T A T V G
R1	Tsr 483–507	V T Q Q N A A L V E E S A A A A A A L E E Q A S R
	Tar 481–505	V T Q Q N A S L V Q E S A A A A A A L E E Q A S R
	Tap 479–503	V T Q Q N A S L V E E A A V A T E Q L A N Q A D R

K1	Tsr 296	E	Q*	Q	A	A	303	E	E*	T	A	A	310	E	Q*	L	T	A
	Tar 294	E	Q	Q	A	S	302	E	E	T	A	A	309	E	Q	L	T	A
	Tap 293	E	Q	Q	A	A							308	E	Q	L	T	A
R1	Tsr 485	Q	Q	N	A	A	492	E	E	S	A	A	502	E	E	Q	A	S
	Tar 483	Q	Q	N	A	S	490	Q	E	S	A	A	500	E	E	Q	A	S
	Tap 482	Q	Q	N	A	S	488	E	E	A	A	V	494	E	Q	L	A	N

[a] Consensus: E/Q E*/Q X A A. The asterisk denotes the residue that was found to be methylated by Khery and Dahlquist (1982a,b).

(Parkinson and Houts, 1982). The consensus sequence, *EEXAA* (glutamic, glutamic, any amino acid, alanine, alanine), may be the site that is required for the binding of the methylesterase or methyltransferase and thus may represent a recognition site for these enzymes so that methylation occurs at the appropriate residue. It is interesting that the spacing between residues that are methylated is

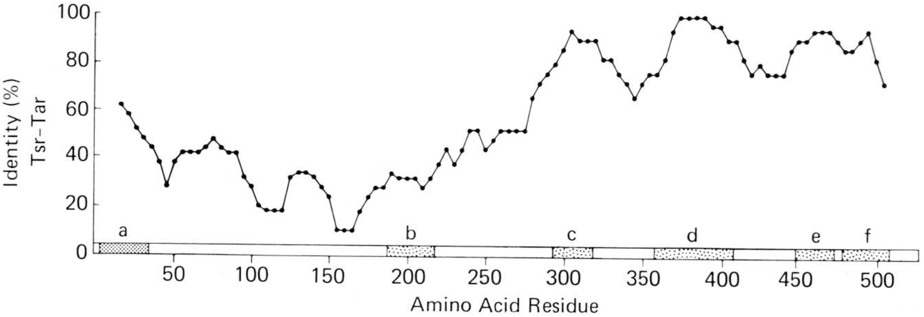

FIG. 4. Tsr–Tar amino acid sequence homology. Shown is the percentage amino acid identity in a window of 29 residues plotted against the central coordinate of that window. The window was moved over the sequence alignment of Fig. 3. For simplicity, a point was plotted for every fifth amino acid. The window size was chosen as that which appeared subjectively to reveal the major features of the relationship between the sequences, while smoothing out much of the noise inherent in such an analysis. Along the abscissa we have indicated the positions of six regions discussed in the text: a, the signal sequence; b, the membrane-spanning domain; c, the methylated peptide K1; d and e, two long, conserved stretches; f, the methylated peptide R1.

relatively constant; there are six or seven amino acids between each of the sites. Thus, for example, in the *tsr* K1 peptide, methylation occurs at residues 297, 304, and 311. If one assumes that the peptides that include the sites of methylation should form α helixes, then sites for methylation are all found to occur on one face of the helix separated from each other by approximately two turns. This conserved orientation may be important in the function of the methylation sites.

In addition to the conserved regions in which methylation occurs, another highly conserved sequence in all three transducers is the amino acid sequence from residue 356 to residue 428 (Fig. 3). It includes 45 amino acids that are 100% indentical in all three proteins. The function of this central conserved sequence is not clear. It could be important for the appropriate binding of the enzymes involved in methylation and demethylation, or it may play some more specific role in generating an excitatory signal.

B. Hydrophobic Sequences

A scan of the amino acid sequences for each of the three polypeptides reveals only two stretches of hydrophobic residues that could act as transmembrane elements. The first occurs within the first 30 amino acids of the proteins. The hydrophobic sequence includes all of the properties that one might expect for a signal peptide that is involved in initiating the process of transmembrane transport of the transducer protein. Thus, for example, the long hydrophobic stretch of approximately 27 amino acids is bounded on each side by a series of charged residues, and is sufficiently long to traverse the membrane (Fig. 3). The second hydrophobic sequence is found at approximately residue 189 in the *tar* sequence and extends to residue 215. This is again found in all of the transducers. There is a stretch of at least 24 amino acids that are all hydrophobic and could again act as membrane-spanning regions that stop the transmembrane transfer of the polypeptide and anchor the protein in a transmembrane configuration.

IV. A MODEL OF THE STRUCTURE OF THE SENSORY TRANSDUCERS

On the basis of the amino acid sequences of these three transducers, it is possible to propose a specific model that accounts for the distribution of the transducer molecule in the cell membrane (Fig. 5). We suggested (Krikos *et al.*, 1983) that the amino acid sequence corresponding to residues 33–190 in all of the proteins makes up the ligand-binding region that is located on the periplasmic surface of the membrane. This stretch of amino acids could fold into a series of ligand-binding sites that could account for the receptor activity of the transducer protein. The highly conserved portion of the protein from residue 215 to the C

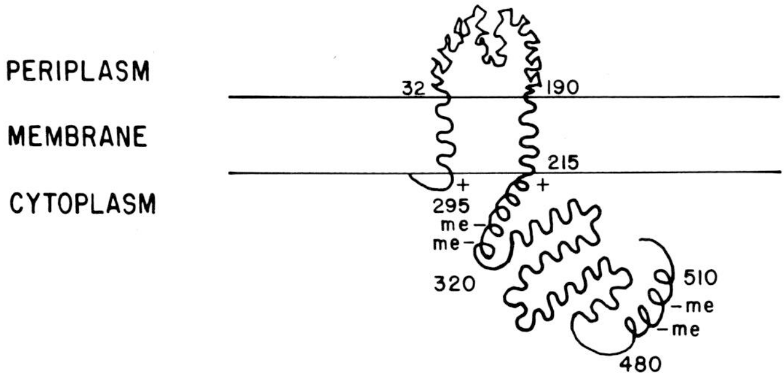

FIG. 5. A model of the transmembrane disposition of the sensory transducer proteins. In this case the numbering refers to the Tsr sequence. The polypeptide is shown spanning the membrane twice, at the amino-terminal signal sequence (which may be proteolytically removed upon assembly), and at the internal stop-transfer sequence. The sites of methylation (me) located at the cytoplasmic side of the membrane are clustered in two α-helical regions. The same overall structure is envisaged for all three transducers considered here.

terminus would be on the cytoplasmic side of the membrane and would include a domain involved in generating an excitatory signal as well as the structural domains required for adaptation. This model is shown in cartoon form in Fig. 5. The model suggests that the transducer functions by binding specific ligands in the periplasmic space. Binding must then result in an alteration in the structure of the transducer, either changing its conformation or its state of aggregation, or both. Somehow this structural change must be transmitted across the membrane so that it affects the cytoplasmic portion of the protein leading to signal formation and adaptation. A similar structural model was proposed by Russo and Koshland (1983) from their determination of the sequence of the *tar* gene in *Salmonella*. More recently, the structure of the *trg* gene product was suggested to be the same as that for all of the other transducers based upon the determination of the sequence of *trg* (Bollinger *et al.*, 1984).

One clear prediction of the structural model that derives from the amino acid sequences of the transducer proteins is that ligand-binding specificity of the receptor is determined primarily by the periplasmic portion of the polypeptide and this region exists as an independently functioning domain of the protein. The cytoplasmic and periplasmic portions of the proteins would have their own discrete structures and they might then be interchangeable. Thus, it should be possible to exchange portions of the *tar* and *tsr* genes so that a chimeric gene could be formed with the N terminus of the *tar* gene and the C terminus of the *tsr* gene. We would predict that such a chimeric gene product might function as a transducer having the ligand-binding specificity of aspartate and using the signal-generating mechanism encoded by the serine-specific transducer gene. As a general paradigm, the formation of chimeric genes using portions of each of the

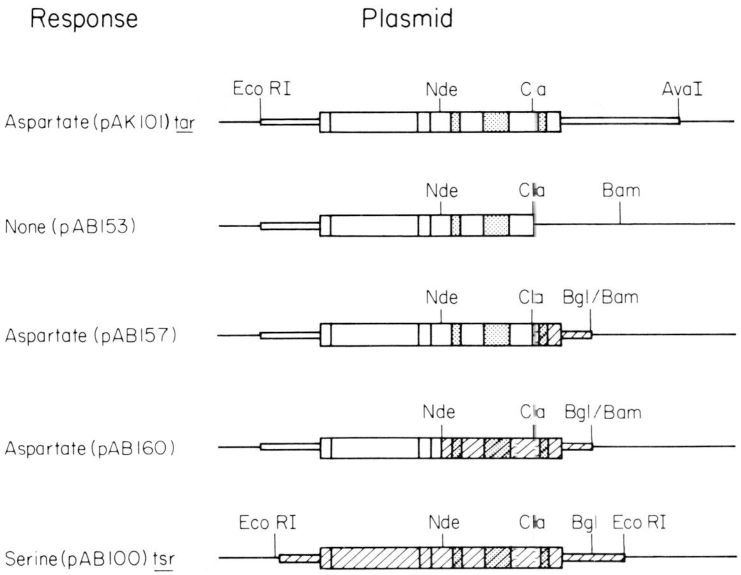

FIG. 6. The construction of chimeric transducer genes. The top line shows the schematic figure of the *tar* gene; the stipled regions represent the part of the gene that determines the constant region and the R1 and K1 peptides. The wide box represents the region that encodes the transducer. The narrower box represents the rest of the insert, and the solid line represents the plasmid. In pAB157 the Cla–Bgl fragment of the *tsr* gene was inserted into the Cla site of pAB153 so that it replaced the missing Cla fragment. The plasmid pAB160 carries a gene made up of the N-terminal half of the *tar* gene and the C-terminal half of the *tsr* gene. The relative activity of each plasmid in tests of chemotaxis is shown in the column at the left.

homologous gene regions could allow one to map the functions of specific domains of the polypeptide. Figure 6 illustrates some of the constructions that were initially prepared to test this approach. First, the *tar* and *tsr* genes were cloned both onto a multicopy plasmid pBR322 and onto a λ phage that could be introduced into a cell as a prophage. This allows us to modify the genes and then reintroduce them in a multicopy form (in the plasmid) or as a single copy gene (in the phage). The genes were introduced into a strain of *E. coli* that had been deleted for the *tar* and *tap* genes and for the *tsr* gene. Thus, this strain was not able to show chemotaxis toward any of the test compounds on motility plates. However, upon introduction of the plasmid-carrying *tar* or *tsr* genes or upon lysogenization with the appropriate bacteriophage, chemotaxis was restored and could be readily seen when the strain was stabbed into tryptone motility plates. A series of constructions was made using conserved restriction sites that were found both in the *tar* and the *tsr* genes (Fig. 6). Plasmid pAB153 represents a deletion of the C-terminal end of the *tar* gene starting at a Cla restriction endonuclease site. This deletion results in the loss of the 84 terminal amino acids of the *tar* gene product. Two new amino acids are added before a termination site is encountered. This plasmid pAB153 was tested for its ability to confer chemotaxis on the triply deleted transducer strain and it had no apparent activity. In order to test whether activity could be restored by adding back genetic material from the corresponding region of the *tsr* gene, a fragment starting at the Cla site derived from the *tsr* gene was inserted to form the chimeric plasmid pAB157. When this plasmid was used in a triply deleted strain, it restored chemotaxis activity and the cells were shown to respond to aspartate.

Another construction was made to test the ability of portions of the gene to restore activity. The plasmid pAB160 was constructed using the conserved NdeI restriction site. This site is found in an homologous spot in both the *tar* and *tsr* genes, and it results in the formation of a polypeptide in which the first 257 amino acids are derived from the *tar* gene and the terminal 279 amino acids are derived from the *tsr* gene. Again, this chimeric plasmid was found to endow the cell with chemotaxis and it showed responses only to aspartate and not to serine. The specificity of the response was measured by introducing the chimeric gene on a plasmid or λ bacteriophage and infecting a host cell in which the *tar, tap,* and *tsr* genes were deleted. The specificity of the response was then followed in three ways: (1) on motility plates that were saturated (10 m*M*) with either aspartic acid or serine, (2) by following the changes in the distribution of the electrophoretic mobility of the polypeptides that were encoded by the chimeric genes after adding either aspartate or serine, and (3) by following rotation of the tethered cells after addition of serine or aspartate. In all cases the specificity of the response to attractant correlated with the origin of the N-terminal portion of the proteins and not with the C-terminal end. These results are consistent with the notion that the ligand-binding specificity of the transducer protein resides in the N-terminal region of the protein and that the adaptation and signaling functions are interchangeable. Thus, we imagine that ligand binding in the periplasmic space affects the structure of the transducer protein and that this effect is propagated and somehow transmitted to the C-terminal region of the protein which is involved in adaptation and in generating a signal that can be transmitted to the flagellar organelle.

This model raises a number of important questions, e.g., how does information concerning ligand binding get transmitted across the membrane? The C-terminal portion of the Tar and Tsr proteins appears to be interchangeable. Is this kind of interchange possible with all of the other transducers? Do different C-terminal regions signal and adapt with different efficiencies? Can other functional subdomains in the molecule be delineated? A variety of other chimeric genes can be readily constructed. They can be designed to probe the relative functional independence of domains of the transducer polypeptide. These experiments may allow us to draw clear conclusions about the mechanisms involved in the function of the receptor–transducer molecules. The receptors in this relatively simple system have many properties that are similar to receptors in more complex systems. Thus, it may be possible to arrive at some generalizations that are applicable to the understanding of the function of receptor–transducers in a variety of contexts.

ACKNOWLEDGMENTS

This work was supported by N.I.H. Grant AI19296 to Dr. M. Simon.

REFERENCES

Adler, J. (1969). Chemoreceptors in bacteria. *Science* **166,** 588–597.

Berg, H. C., and Anderson, R. A. (1973). Bacteria swim by rotating their flagella filaments. *Nature (London)* **245,** 380–382.

Berg, H. C. (1975). Chemotaxis in bacteria. *Annu. Rev. Biophys. Bioeng.* **4,** 119–139.

Bollinger, J., Park, C., Harayama, S., and Hazelbauer, G. (1984). Structure of the *trg* protein. *Proc. Natl. Acad. Sci. U.S.A.*, in press.

Boyd, A., and Simon, M. (1982). Bacterial chemotaxis. *Annu. Rev. Physiol.* **44,** 501–517.

Boyd, A., Krikos, A., and Simon, M. (1981). Sensory transducers of *E. coli* are encoded by homologous genes. *Cell* **26,** 333–343.

Boyd, A., Kendall, K., and Simon, M. (1983). Structure of the serine chemoreceptor in *E. coli. Nature (London)* **301,** 623–626.

Clarke, S., and Koshland, D. E. (1979). Membrane receptors for aspartate and serine in bacterial chemotaxis. *J. Biol. Chem.* **254,** 9695–9702.

Hazelbauer, G. L., and Engstrom, P. (1981). Multiple forms of methyl accepting chemotaxis proteins. *J. Bacteriol.* **145,** 35–42.

Hazelbauer, G. L., and Parkinson, J. S. (1977). *In* "Receptors and Recognition: Microbiological Interactions" (J. Reissing, ed.), Ser. B, Vol. 3, pp. 59–98. Chapman & Hall, London.

Hedblom, M. L., and Adler, J. (1980). Genetic and biochemical properties of *E. coli* mutants with defects in serine chemotaxis. *J. Bacteriol.* **144,** 1048–1060.

Khery, M. R., and Dahlquist, T. W. (1982a). Adaptation in bacterial chemotaxis: cheB-dependent modification permits additional methylation of sensory transducer proteins. *Cell* **29,** 761–772.

Khery, M. R., and Dahlquist, T. W. (1982b). The methyl accepting chemotaxis proteins of *E. coli:* Identification of multiple methylation sites on MCPI. *J. Biol. Chem.* **257,** 10378–10386.

Koshland, D. E. (1979). A model regulatory system: Bacterial chemotaxis. *Physiol. Rev.* **59,** 811–862.

Koshland, D. E. (1981). Biochemistry of sensing and adaptation in a simple bacterial system. *Annu. Rev. Biochem.* **50,** 765–782.

Krikos, A., Mutoh, N., Boyd, A., and Simon, M. (1983). Sensory transducers of *E. coli* are composed of discrete structural and functional domains. *Cell* **33,** 615–622.

Larsen, S. H., Reader, R. W., Kort, E. M., Tso, W. W., and Adler, J. (1974). Change in direction of flagellar rotation is the basis of the chemotactic response in *E. coli. Nature (London)* **249,** 74–77.

Macnab, R. M. (1978). Bacterial motility and chemotaxis: The molecular biology of a behavioral system. *C.R.C. Crit. Rev. Biochem.* **5,** 291–341.

Macnab, R. M., and Koshland, D. E. (1972). A gradient sensing mechanism in bacterial chemotaxis. *Proc. Natl. Acad. Sci. U.S.A.* **69,** 2509–2512.

Parkinson, J. S. (1981). Genetics of bacterial chemotaxis. *In* "Genetics as a Tool in Microbiology"

(S. W. Glover and D. A. Hopwood, eds.), pp. 265–290. Cambridge Univ. Press, London and New York.

Parkinson, J. S., and Houts, S. E. (1982). Isolation and behavior of *E. coli* deletion mutants lacking chemotaxis functions. *J. Bacteriol.* **151,** 106–113.

Russo, A. T., and Koshland, D. E. (1983). Separation of the signal transduction and adaptation functions of the aspartate receptor in bacterial sensing. *Science* **220,** 1016–1020.

Silverman, M., and Simon, M. (1974). Flagellar rotation and the mechanism of bacterial motility. *Nature (London)* **249,** 73–74.

Silverman, M., and Simon, M. (1977a). Chemotaxis in *E. coli:* Methylation of *che* gene products. *Proc. Natl. Acad. Sci. U.S.A.* **74,** 3317–3321.

Silverman, M., and Simon, M. (1977b). Bacterial flagella. *Annu. Rev. Microbiol.* **31,** 397–419.

Springer, M. S., Goy, M. T., and Adler, J. (1977). Sensory transduction in *E. coli:* Two complementary pathways of information processing. *Proc. Natl. Acad. Sci. U.S.A.* **74,** 3312–3316.

Springer, M. S., Goy, M. T., and Adler, J. (1979). Protein methylation in behavioural control mechanisms and in signal transduction. *Nature (London)* **264,** 577–579.

Wang, E. A., and Koshland, D. E. (1980). Receptor structure in the bacterial sensing system. *Proc. Natl. Acad. Sci. U.S.A.* **77,** 7157–7161.

Chapter 2

Mutational Analysis of the Structure and Function of the Influenza Virus Hemagglutinin

MARY-JANE GETHING, CAROLYN DOYLE, MICHAEL ROTH, AND JOE SAMBROOK

Cold Spring Harbor Laboratory
Cold Spring Harbor, New York

I. INTRODUCTION

In recent years the amino acid sequences of a number of integral membrane proteins have been determined, either using classical protein chemistry techniques or, indirectly, by analysis of the DNA sequences of cloned copies of their genes. These proteins include those found naturally on eukaryotic cell membranes as well as virus-coded polypeptides whose biosynthesis utilizes host cell

ISBN 0-12-153323-9

processes and enzymes for their translation, membrane translocation, glycosylation, maturation, and transport to the cell surface.

The hemagglutinin (HA) glycoprotein of influenza virus is the best characterized of all integral membrane proteins. The complete amino acid sequences of HAs from a number of influenza virus strains are known, and the genes encoding these HAs have been cloned and sequenced (for a review see Lamb, 1983). The three-dimensional structure of the HA molecule (Wilson *et al.*, 1981), its organization into trimers (Wiley *et al.*, 1977), and its orientation with respect to the membrane (Skehel and Waterfield, 1975) have been determined. In addition, the points at which the protein is glycosylated (Gething *et al.*, 1980; Wilson *et al.*, 1981) and the location of the major antigenic sites on the molecule (Wiley *et al.*, 1981) have been defined. This wealth of knowledge makes the HA an ideal candidate for site-specific mutagenesis experiments aimed at elucidating the function of the various domains of the molecule. Apart from the specialized features associated with its antigenicity and biological roles of attachment (Hirst, 1941) and penetration (Matlin *et al.*, 1981; White *et al.*, 1982), the structure of the HA molecule is characteristic of the major class of cellular integral membrane proteins, having typical amino- and carboxy-terminal hydrophobic regions. Thus the HA provides a useful model system for the study of the structure, function, and biosynthesis of eukaryotic membrane proteins in general.

The crucial interactions between HA and membranes are mediated by three separate hydrophobic regions of the molecule. Two of these—the signal sequence that draws the nascent polypeptide chain into contact with the intracellular membrane system (Blobel *et al.*, 1979; Davis and Tai, 1980; Walter and Blobel, 1981) and the carboxy-terminal hydrophobic sequence that anchors the completed molecule in the lipid bilayer—are not highly conserved in amino acid sequence between HAs from different strains of virus (Gething *et al.*, 1980; Lamb, 1983). This suggests that hydrophobicity rather than specific sequence is important in the interaction of the amino and carboxy termini with lipid bilayers. In contrast, the third hydrophobic sequence, which lies at the N terminus of the HA2 subunit, is highly conserved among different HAs (Air, 1981). This region is involved in the low pH-induced fusion of viral and cellular membranes that marks the onset of infection of the host cell by the virus (Gething *et al.*, 1978; Richardson *et al.*, 1980; Garten *et al.*, 1981; White *et al.*, 1982). The N-terminal signal, the C-terminal anchor, and the cytoplasmic tail domains of HA are characteristic of those found in the majority of integral membrane proteins. Thus it would be of general interest to determine the effects of alterations in these regions on the biosynthesis, structure, and transport of the molecule.

The HA protein is naturally encoded by an RNA genome that is not amenable to *in vitro* mutagenesis. However, the HA gene from the A/Japan/305/57 strain of influenza virus has now been converted to double-stranded DNA, cloned, sequenced, and inserted into a variety of prokaryotic and eukaryotic expression vectors (Gething *et al.*, 1980; Heiland and Gething, 1981; Gething and Sambrook, 1981, 1983). When the HA cDNA is introduced into eukaryotic cells using vectors derived from SV40, it is expressed with high efficiency into a fully glycosylated protein that is displayed on the infected cell's surface in an antigenically and biologically active form. The structure and properties of this glycoprotein are indistinguishable from those of authentic HA from influenza virus-infected cells (Gething and Sambrook, 1981). These cloned, expressing copies of the HA gene provide the material to analyze the effect of mutations on the structure and function of the protein.

II. EXPRESSION OF WILD-TYPE HA IN SIMIAN CELLS USING A RECOMBINANT SV40 VIRAL VECTOR

Because the HA gene occurs naturally in the form of negative-stranded RNA, the cloned DNA copy contains none of the controlling elements that are required for efficient transcription of conventional DNA genes. Thus the SV40 vector shown in Fig. 1 has been designed so that the HA coding region is placed under the control of the SV40 late promoter(s) replacing the sequences coding for the SV40 capsid genes (Gething and Sambrook, 1981). The SV40 genome contains

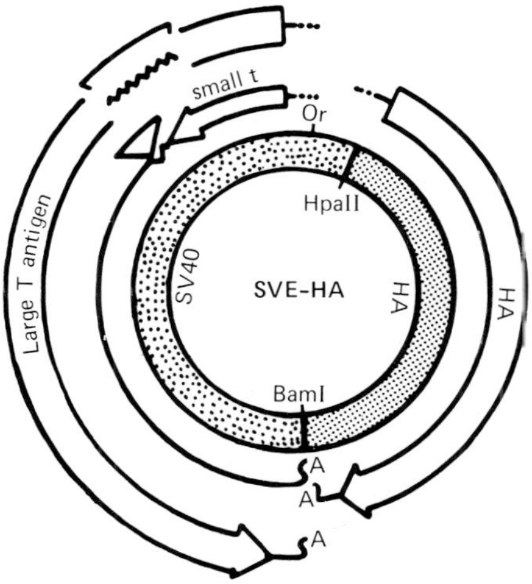

FIG. 1. SV40-HA recombinant vector designed to express hemagglutinin in simian cells The shaded circle represents the double-stranded circular DNA genome. The protein-coding sequences are indicated by blocked arrows; the zigzag line represents the sequence spliced out; a wavy line with an A represents the 3′-terminal poly(A) tract. For details of the construction of the vector see Gething and Sambrook (1981).

an enhancer sequence close to the origin of replication (Tooze, 1980). In addition, to ensure that transcripts of the HA gene would be terminated efficiently, the HA sequences were inserted upstream of a poly(A)-addition signal in the SV40 genome.

A recombinant SV40 virus will replicate efficiently in eukaryotic cells only if (1) the host cells are permissive for viral DNA synthesis, (2) the vector contains a functional origin of replication, (3) there is a supply of large T antigen sufficient to initiate successive rounds of SV40 DNA synthesis, and (4) a supply of the SV40 capsid proteins VP1, VP2, and VP3 is available for production of infectious virions containing the recombinant genome (Gluzman, 1981; Tooze, 1980). Because the SV40-HA recombinant contains a functional origin of DNA replication and an intact set of SV40 early genes coding for T antigen, it will replicate efficiently in permissive simian cells. However the late genes of SV40 have been deleted, and production of infectious virions containing the recombinant genome therefore requires that SV40 capsid proteins be supplied by a complementing helper virus—an SV40 mutant [d11055 (Pipas *et al.*, 1979)] that carries a defect in the early region of the viral genome. Finally, efficient packaging of recombinant genomes into SV40 capsids requires that the genome size remain within 70–105% of that of the wild-type virus (Tooze, 1980). When the HA gene is inserted into the SV40 genome between the *Hpa*II site at nucleotide 346 and the *Bam*HI site at nucleotide 2533 (Fig. 1), this size restraint is accommodated.

The SV40-HA recombinant was introduced into simian CV-1 cells together with an equal amount of the DNA of the SV40 deletion mutant d11055, using DEAE–dextran as facilitator (McCutchan and Pagano, 1968). At any one time only a small proportion of cultured mammalian cells are competent to take up DNA, so that initially very few cells of the population are infected. However, during the next several days, each of these cells undergoes a lytic infection and produces over a million virus particles that are able to spread into neighboring cells and infect them with high efficiency. Usually the lysate obtained from the first set of infected cells needs to be passed serially once or twice to obtain a high-titer virus stock. This stock consists of approximately equal numbers of helper virus particles and recombinants and can be used to infect permissive cells with high efficiency and to induce in them lytic cycles of viral growth. During such infections the viral genomes are transported to the nucleus where they are liberated from their capsids. The early promoter soon becomes active and the T antigen genes are expressed. By 12 hr after infection, viral DNA replication is under way. After approximately 20 hr the HA gene in the recombinant genome and the capsid genes in the helper genome begin to be expressed with high efficiency from their respective late promoters. By 36–48 hr after infection, the newly synthesized viral genomes begin to be assembled into progeny virus particles—a process that continues for another 24 hr, whereupon the cells detach from their substrate and die. Thus it is possible to analyze the expression of HA in the infected cells between 20 and 65 hr postinfection.

To establish that the protein expressed from the recombinant vector was authentic in its structure, antigenicity, and biological activities, a series of assays was performed including radioimmunoassay, immunoprecipitation, immunofluorescence, and hemagglutination and cell fusion assays. These experiments have been described in detail previously (Gething and Sambrook, 1981; White *et al.*, 1982) and will be discussed only briefly here. Similar results have been reported for the expression from SV40 recombinant vectors of HAs from other strains of influenza virus (Sveda and Lai, 1981; Hartman *et al.*, 1982).

1. Radioimmunoassay

The amounts of Japan HA expressed from the recombinant have been quantitated using a solid-phase radioimmunoassay (Gething and Sambrook, 1981). It was found that the amount of HA detected increased as the course of the infection proceeded. By 62 hr, when the lytic infection was in its terminal phase, cells contained approximately 6×10^8 molecules of HA per cell, i.e., about 200 μg HA per 10 cm dish of cells. For comparison a simian cell at a late stage during infection with influenza virus contains about 5×10^7 molecules of HA per cell.

2. Immunoprecipitation

Extracts of cells infected with the recombinant virus and labeled either with [^{35}S]methionine or with radioactive sugar precursors contained a protein that was specifically precipitated by anti-HA sera. The protein was indistinguishable in size from authentic glycosylated HA precipitated from extracts of cells infected with influenza virus. These cells contained so much of the protein that it was not necessary to use immunoprecipitation to detect it; HA could be seen either as a band stained with Coomassie blue after the extracts of infected cells had been analyzed by SDS–PAGE or as a prominent radioactive species when extracts were prepared from infected cells labeled with [^{35}S]methionine. Treatment of the infected cells with tunicamycin resulted in the production of a smaller nonglycosylated protein that was identical in size to nonglycosylated authentic HA.

3. Immunofluorescence

Cells infected with the SV40-HA recombinant, fixed and stained for cytoplasmic fluorescence using anti-HA serum, displayed bright, perinuclear fluorescence with a region corresponding to the Golgi apparatus staining with particular intensity. The surface of the cells also stained specifically with a uniform dimmer fluorescence. The distribution of fluorescence in cells infected with influenza virus was similar to that displayed by recombinant infected cells, but of lower intensity.

4. Hemagglutination and Erythrocyte Binding to Infected Cells

Guinea pig red blood cells could be agglutinated by extracts of cells infected with the SV40 HA virus. Intact monolayers of infected cells absorbed a dense carpet of erythrocytes onto their surfaces.

5. Cell–Cell Fusion

CV-1 cells infected with the recombinant virus could be fused to form giant polykaryons if the monolayers were first treated with low levels of trypsin to cleave HA0 to HA1 and HA2 and then exposed to transient low pH (White *et al.*, 1982). The pH threshold at which the fusion occurs is identical to that observed for the low pH-induced fusion of cells infected with the Japan strain of influenza virus.

III. ANALYSIS OF THE EXPRESSION OF MUTANT HA PROTEINS

From the studies described above, it is apparent that the HA protein that is normally encoded by a negative-stranded RNA genome can be expressed in copious amounts when double-stranded DNA copies of its coding sequences are harnessed to a strong SV40 promoter. The HA expressed from the wild-type gene appears normal in all respects; its molecular weight is indistinguishable from that of the authentic protein, and it is displayed on the infected cell's surface in a glycosylated form that is both biologically and antigenically active. It is therefore feasible to introduce mutations into the cloned HA nucleotide sequence, to express the altered genes in eukaryotic cells, and to analyze the phenotype of the mutant proteins. In Fig. 2 a linear map of the A/Japan/305/57 HA gene relates the various functional domains of the molecule to the major restriction sites in the nucleotide sequence. Below we describe experiments in which sequences coding for the hydrophobic domains and the cytoplasmic tail of the wild-type HA molecule have been deleted, altered, or exchanged for those of other eukaryotic membrane proteins.

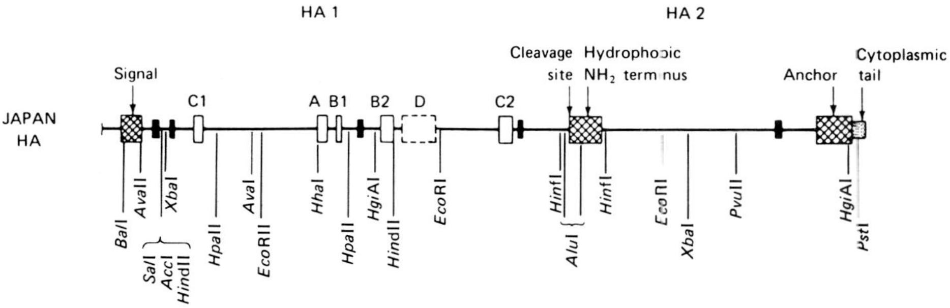

FIG. 2. Linear map of HA from the A/Japan/305/57 strain of influenza virus. The map relates the various functional regions of the molecule to the restriction sites in the gene. ▩, Hydrophobic sequence; ■, glycosylation site; □, antigenic site.

A. Deletion of the Signal Sequence Converts HA to a Nonglycosylated, Intracellular Protein

The double-stranded exonuclease *Bal*31 was used to remove from the wild-type HA sequences the 5′ nucleotides coding for the N-terminal signal peptide. The truncated gene was used to replace the wild-type gene in the SV40-HA vector, so that the HA initiation codon was fused in frame to the codon for the first amino acid of the mature HA polypeptide, leaving unaltered the remainder of the coding sequences (Gething and Sambrook, 1982).

A variant, unglycosylated HA protein was detected by immunoprecipitating extracts of cells infected with a virus stock containing the mutant recombinant (Gething and Sambrook, 1982). Its molecular weight (M_r = 61,000) was identical to that of nonglycosylated HA synthesized either *in vivo* in the presence of tunicamycin (Fig. 3) or *in vitro* by translation in a reticulocyte lysate. This result is consistent with the signal-minus HA being synthesized as a cytoplasmic protein on free polyribosomes. Because it lacks a signal sequence, the nascent polypeptide is not translocated through the membrane of the endoplasmic reticulum. It is therefore never exposed to the glycosylating enzymes that reside on the luminal side of the membrane. The signal-minus mutant is expressed at a relatively low level (10^6 molecules/cell) compared to the wild-type (6×10^8 molecules/cell). Control experiments suggested that this poor yield stems from a combination of factors including instability of both the mutant mRNA and the nonglycosylated variant protein.

These results clearly confirm that the signal sequence is required for correct translocation of the nascent polypeptide. It is now feasible to construct and analyze HA mutants containing single nucleotide changes in the DNA sequence encoding the hydrophobic peptide. Such experiments should elucidate the role of

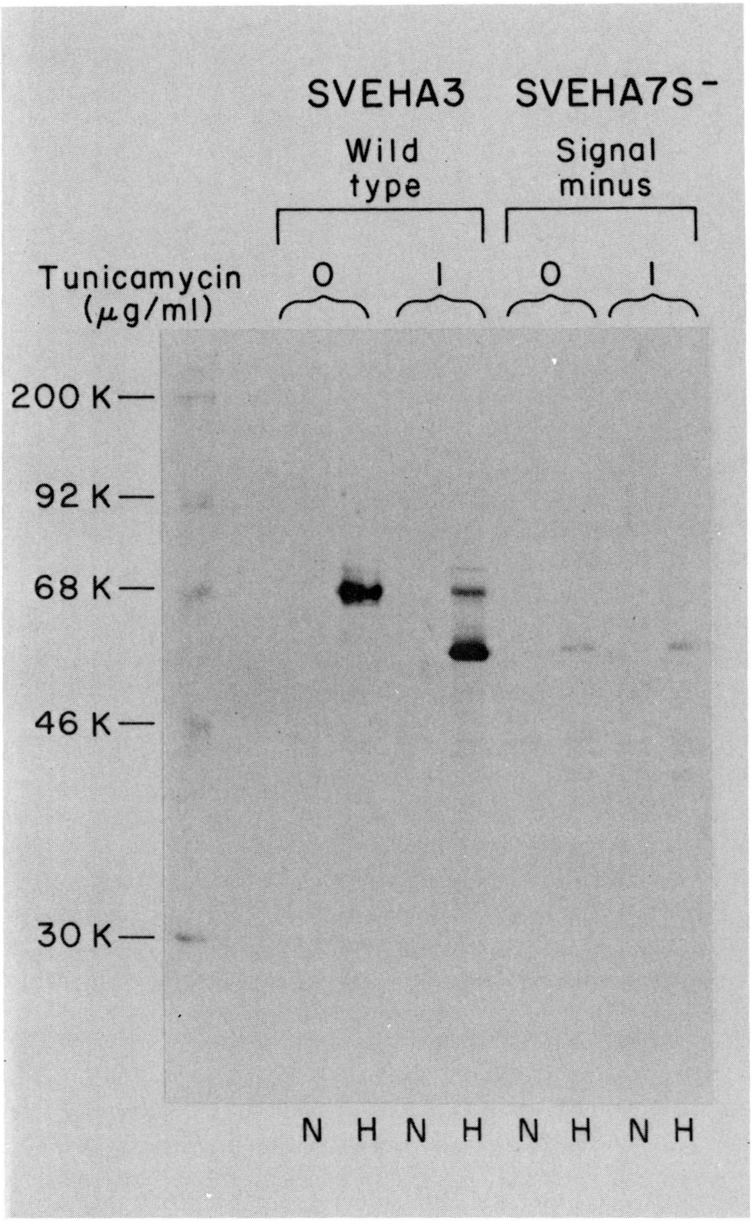

FIG. 3. Expression of wild-type and anchor-minus HAs in CV-1 cells: effect of tunicamycin on the molecular weight of the protein. Monolayers of CV-1 cells, infected with the recombinant genomes as described previously (Gething and Sambrook, 1981), were treated with tunicamycin (0 or 1 μg ml^{-1} in DME containing 10% serum) between 46 and 48 hr after infection. The cells were then labeled with [^{35}S]methionine in the presence of the same concentration of tunicamycin. Cell extracts were prepared, immunoprecipitated, separated by SDS–PAGE, and autoradiographed.

particular amino acids in signal function and determine the effect of the nucleotide sequence in this region on the stability and efficiency of translation of the HA mRNA.

B. Deletion of Sequences Coding for the C-Terminal Hydrophobic Region Converts HA into a Secreted Protein

During its biosynthesis the wild-type HA becomes anchored into the lipid bilayer of the endoplasmic reticulum by a stretch of approximately 24–27 amino acids that are coded by nucleotide sequences near the 3′ end of the HA gene (Gething *et al.,* 1980). The construction of an anchor-minus mutant utilized a plasmid (p2G10) that was originally cloned lacking sequences corresponding to the C-terminal anchor and cytoplasmic tail (Gething *et al.,* 1980). The truncated gene replaced the wild-type HA gene in the original SV40-HA recombinant to create a new recombinant that codes for a protein that lacks the 38 amino acids normally found at the C terminus of HA (Gething and Sambrook, 1982). Substituting for them is a stretch of 11 amino acids, largely polar in nature, that are encoded by the dG:dC homopolymer tail, a synthetic *Bam*HI linker, and a short sequence of SV40 DNA.

The anchor-minus mutant HA is synthesized very efficiently in CV-1 cells infected with the mutant recombinant (1×10^9 molecules/cell). However, instead of being entirely cell-associated (in the Golgi or on the cell surface), the mutant HA is secreted efficiently into the medium (Fig. 4) (Gething and Sambrook, 1982). It is fully glycosylated and, like the wild-type protein (Wiley *et al.,* 1977), the mutant HA is assembled as a trimer of HA subunits. The reduction of approximately 3000 Da in its apparent molecular weight is consistent with the extent of deletion in the gene. Thus removal of the carboxy-terminal sequences results in the loss of anchoring function. The nascent polypeptide, instead of remaining attached to the luminal face of the rough endoplasmic reticulum, passes completely through the membrane. Once free in the lumen, the mutant HA is treated by the cell as if it were an authentic secretory protein and is discharged into the medium.

The secreted form of HA appears to differ from the membrane-bound wild-type protein in its rate of transport and terminal glycosylation. In eukaryotic cells the addition of carbohydrate to proteins takes place in two stages. As the nascent polypeptide appears on the luminal side of the rough endoplasmic reticulum, preformed mannose-rich oligosaccharides are transferred from a lipid-dolichol carrier to certain asparagine residues marked by the amino acid sequence Asn–X–Ser or –Thr (Hubbard and Robbins, 1979; Neuberger *et al.,* 1972). This co-translational ''core'' glycosylation is the first step in an elaborate program of

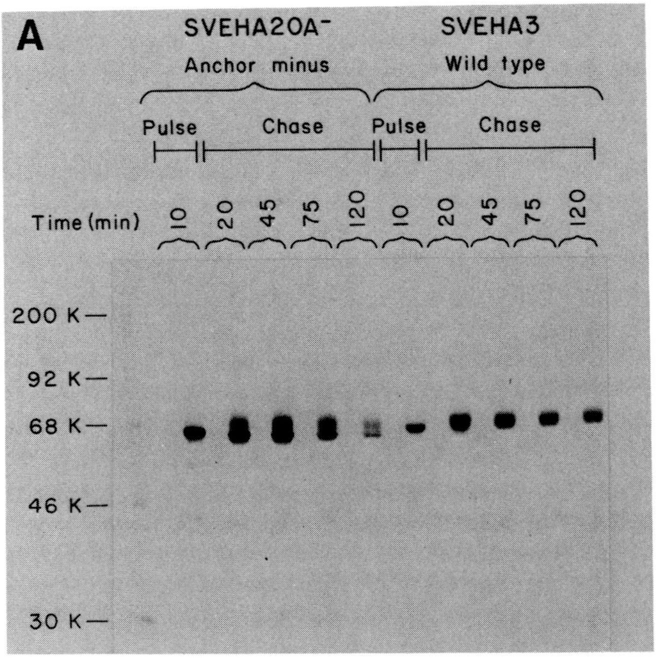
A
SVEHA20A⁻
Anchor minus
SVEHA3
Wild type
Pulse
Chase
Pulse
Chase
Time (min)
10
20
45
75
120
10
20
45
75
120
200 K
92 K
68 K
46 K
30 K

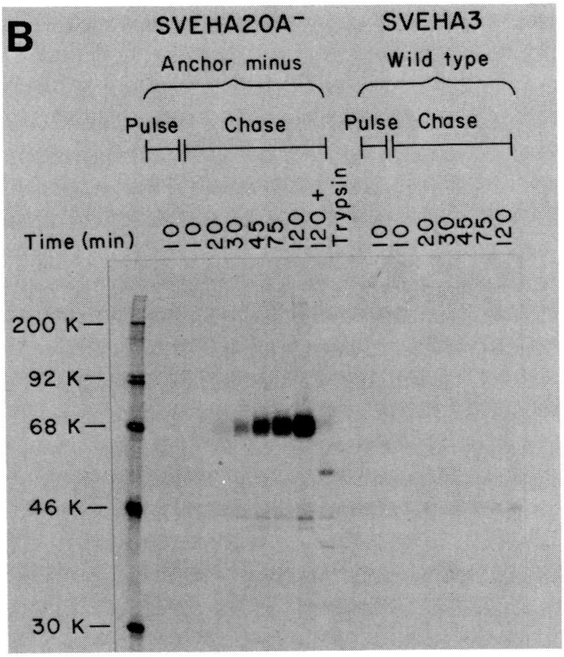
B
SVEHA20A⁻
Anchor minus
SVEHA3
Wild type
Pulse
Chase
Pulse
Chase
Time (min)
10
10
20
30
45
75
120
120 + Trypsin
10
10
20
30
45
75
120
200 K
92 K
68 K
46 K
30 K

reactions that take place initially in the rough endoplasmic reticulum and later in the Golgi apparatus, in which sugar residues are removed from and added to the newly synthesized protein (Hubbard and Robbins, 1979; Tabas and Kornfeld, 1979). The mannose-rich core oligosaccharides can be removed from the protein backbone by *in vitro* treatment with the enzyme endoglycosidase H (endo H) (Tarentino and Maley, 1974). However, the enzyme does not recognize side chains that have been trimmed or extended. Thus the acquisition of resistance to digestion by endo H can be used to follow the movement of the glycoprotein from the endoplasmic reticulum to the Golgi apparatus. In addition, during the biosynthesis of HA, the transition from core to terminal glycosylation can be visualized as a molecular weight increase on analysis by SDS–PAGE of protein labeled in pulse–chase experiments with [^{35}S]methionine or tritiated sugar precursors (Fig. 4A). Such experiments indicated that, in contrast with the wild-type membrane-bound HA that is glycosylated rapidly and relatively synchronously, the population of secretory molecules becomes terminally glycosylated over a very protracted period.

In summary, the hydrophobic region at the C terminus of HA is required to anchor the protein in cell membranes. Deletion of this region converts HA from an integral membrane protein into a secretory protein. This suggests that signal sequences apart, there is no obligatory sequence of amino acids common to all secretory proteins that causes them to be actively transported out of the lumen of the endoplasmic reticulum and secreted from the cell. Similar results have been obtained using an anchor-minus mutant of an HA from another strain (A/Udorn/72) of influenza virus (Sveda *et al.*, 1982) and an anchor-minus mutant of the G glycoprotein from vesicular stomatitis virus (VSV) (Rose and Bergman, 1982).

C. Anchoring and Cytoplasmic Domains Can Be Exchanged between Integral Membrane Proteins

To determine whether hydrophobic anchoring sequences and cytoplasmic domains can be functionally interchanged between different membrane glycoproteins, we have constructed genes coding for chimeric proteins (Figs. 5 and 6). In the first experiment, *Bam*HI linkers were ligated to a fragment of herpesvirus DNA (Frink *et al.*, 1983) that contains sequences coding for the hydrophobic

---

FIG. 4. Expression of wild-type and anchor-minus HAs in CV-1 cells. Monolayers of CV-1 cells were infected with the recombinant viruses as described previously (Gething and Sambrook, 1981). Forty-eight hours later the cells were labeled with [^{35}S]methionine for 10 min. After two washes with complete medium the cells were incubated for further periods of time as shown in the figure. At each time point the supernatant medium was collected and cell extracts were prepared. (A) Aliquots of the cell extracts were immunoprecipitated using rabbit nonimmune serum (first track of each pair) or anti-Ha serum (second track of pair), separated by SDS–PAGE and autoradiographed. (B) Aliquots of the supernatant media were separated by SDS–PAGE without prior precipitation.

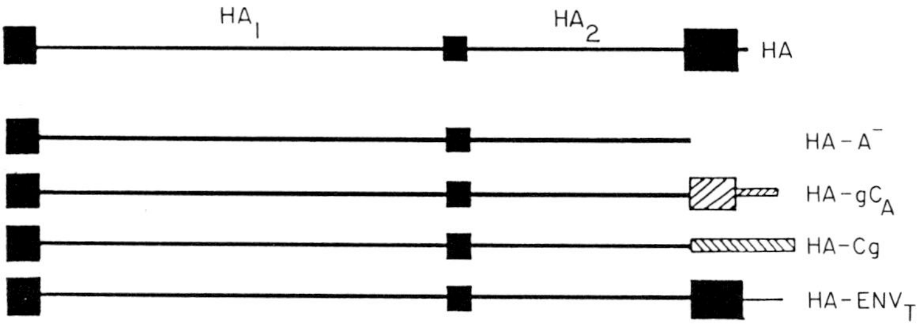

FIG. 5. Schematic representation of the structures of the wild-type, anchor-minus, and chimeric HA proteins. The filled blocks indicate the hydrophobic regions of the HA molecule.

anchor and cytoplasmic tail of the herpes glycoprotein gpC. This fragment was then inserted in both possible orientations into the *Bam*HI site of the SV40-HA recombinant vector following the sequences coding for the truncated, anchor-minus HA. In one orientation (HA-gC) the open reading frame of HA was continued into the new sequences such that the authentic anchor and tail of gpC replaced those of wild-type HA. In the opposite orientation (HA-Cg), another open reading frame (not thought to be utilized in the herpesvirus genome) is fused to the HA sequences such that the carboxy terminus of the truncated protein is extended by 67 amino acids (mostly polar in nature). In a second experiment, a chimeric gene was constructed in which the sequences coding for the cytoplasmic tail of HA were replaced by the analogous sequences from the envelope glycoprotein gene of Rous sarcoma virus (RSV). This construction utilized a mutant HA gene in which a *Cla*I restriction linker had been inserted between the sequences coding for the anchor and tail domains (see Section III,D). The nucleotide sequence between this *Cla*I site and the *Bam*HI site at the end of the HA gene was replaced using a *Taq*I–*Bam*HI restriction fragment which encoded the cytoplasmic tail of the RSV *env* gene (Schwartz *et al.*, 1983), taking advantage of the fact that *Cla*I and *Taq*I restriction enzymes generate compatible ''sticky'' ends.

The various chimeric genes were inserted into the SV40-HA expression vector, and recombinant virus stocks were generated in CV-1 cells All of the chimeric proteins were expressed with high efficiency, comparable to that of the wild-type HA. The cellular location of the proteins was examined by radioimmunoassay, indirect immunofluorescence, and erythrocyte binding (results not shown). Table I compares the time course of transport and the final destination of the chimera with those of the wild-type and anchor-minus HAs. It is immediately obvious that functional anchor and cytoplasmic domains are interchangeable between these glycoproteins. The substitution of the RSV env tail (23 amino acids) for the HA tail (10 amino acids) had no effect on the biosynthesis, rate of transport, or final location of the protein. In particular, the final composition of the carbohydrate side chains appeared to be identical to that of the wild-type HA.

HA wild-type	......KLSSMGVYQILAIYATVAGSLSLAIMMAGISFWMCSNGSLQCRICI
HA anchor-minus	......KLSSMGVYPPPGSRHDKIH
HA-gC anchor	......KLSSMGVYPPPGSEWVGIGIGVLAAGVLVVTAIVYVVRTSQSRQRHRR
HA-Cg	......KLSSMGVYPPPGSQDMHPRLNSCHPLSHGESAHEAREGTRTNAAMITTPPMMVPTPTQAAAGSRARMGGVPGPRGR
HA-env tail	......KLSSMGVYQILAIYATVAGSLSLAIMMAGHRKMINSSINYHTGYRKMQGGAV

FIG. 6. Carboxy-terminal amino acid sequences of the wild-type, anchor-minus, and chimeric HA proteins. The amino acids are shown in the single-letter code. The hydrophobic anchoring domains are boxed.

TABLE I
CELLULAR TRANSPORT OF CHIMERIC HA MOLECULES

	Translocation to ER (signal cleavage) (core glycosylation)	Transport to Golgi: (terminal glycosylation) [Endo H resistance ($t_{\frac{1}{2}}$)]	Transport to surface[a]
HA wild-type	+	+ (20 min)	PM
HA anchor-minus	+	+ (1 hr)	S
HA-gC_{AT}	+	+ (1 hr)	PM
HA-Cg	+	+ (1 hr)	S
HA-ENV_T	+	+ (20 min)	PM

[a] PM, plasma membrane; S, secreted.

The fact that the cytoplasmic tails are exchangeable does not mean that this portion of the molecule plays no role in the biosynthesis or transport of integral membrane proteins. The results presented in Section III,D show that alterations in this region can have significant effects on the transport of HA.

The addition of the anchor and tail of the herpes gpC to the anchor-minus HA restored the interaction with the lipid bilayer so that the chimeric glycoprotein was expressed at the cell surface as an integral membrane protein. On the other hand, the addition of the long polar extension did not interfere with the secretion of the anchor-minus HA. It is of interest, however, that both chimeras retained the mutant phenotype of delayed transport between the endoplasmic reticulum and Golgi apparatus.

We have suggested previously (Gething and Sambrook, 1982, 1983) that the slow transport of the anchor-minus mutant might reflect differences in the efficiency with which membrane-bound proteins and luminal proteins are sequestered into the transport vesicles that travel from the rough endoplasmic reticulum to the Golgi apparatus (Strous and Lodish, 1980). However, the fact that the HA-gC chimera is also slowly transported despite being membrane bound raises the possibility of a signal on the HA molecule that is involved in efficient recognition of the protein for transport between the organelles. Several models have been proposed to describe how the transport of glycoproteins from the endoplasmic reticulum might be facilitated or controlled (Sabatini *et al.*, 1982). One invokes a receptor molecule to which the nascent glycoprotein must bind before it can cross a transport "barrier" between the endoplasmic reticulum and the Golgi. This model, as formulated by Fitting and Kabat (1982), can predict the type of kinetics that we have observed for both the wild-type and mutant HAs and suggests that the wild-type molecule has a high affinity and that the mutants have low affinities for the putative receptor. A second model proposes that newly synthesized proteins passively flow to the Golgi unless a specific interaction with a receptor causes them to be retained or delayed within the endoplasmic reticulum. A further model suggests that there may be no receptor but rather a requirement for the nascent protein to fold correctly or form a multimer (e.g., a trimer in the case of HA) before it can move freely with the lipid flow in the plane of the membrane. Alterations in the protein structure could decrease the probability of forming the correct structure and thus delay exit from the endoplasmic reticulum. Although this model can describe the mechanism of the initial delay in the endoplasmic reticulum, it cannot be used to explain a subsequent block in the Golgi (see below) or the selection between alternative transport pathways, i.e., from the Golgi to the plasma membrane or to the lysosomes. In any case, the requirement for correct folding can be viewed as a prerequisite for transport by either of the first two models.

Neither of the models which invoke a receptor predicts the location of the putative transport signal on the glycoprotein, i.e., whether it might be on the external domain, within the transmembrane region, or in the cytoplasmic tail. Neither do the experiments described above define a location since it is possible either that the signal has been removed along with the carboxy-terminal sequences or that the large deletion (and/or addition) has perturbed a signal that resides on the external domain. Further experiments in which smaller mutations are introduced into the HA sequence will be necessary to confirm and locate such a signal.

In the studies described above, we have analyzed the transport of wild-type and chimeric glycoproteins between the endoplasmic reticulum and the surface of cells that do not have differentiated apical and basolateral plasma membranes. More recently, it has been shown that when the wild-type HA gene is expressed from the SV40-HA recombinant in polarized epithelial cells, the HA is expressed only on the apical surface of the cells (Roth *et al.*, 1983; E. Rodriguez-Boulan, personal communication). It will now be possible, using the chimera described above and others made between HA and glycoproteins normally expressed on basolateral membranes, to determine the location of the sequences that direct different proteins from the Golgi apparatus to the various membranes of eukaryotic cells.

D. Effect of Alterations in the Cytoplasmic Tail on the Intracellular Transport of HA

The major class of eukaryotic integral membrane proteins characteristically is composed of a large external domain, a transmembrane domain of 20–30 amino acids, and a short C-terminal peptide that extends from the cytoplasmic face of the membrane. These cytoplasmic tails are not conserved in sequence between different membrane proteins. They are generally hydrophilic in nature and can vary in length from 2 to 100 amino acids. The tail is exposed to the cytoplasm during transport from the site of synthesis of the protein in the endoplasmic reticulum, via the Golgi apparatus, to its final destination in the membrane of an intracellular organelle or at the cell surface. Thus, it is possible that this region might contain sequences that interact with host cell proteins involved in controlling and facilitating intracellular transport. To investigate whether the cytoplasmic tail of the membrane-bound form of HA might contain such a recognition signal, mutations have been introduced into the nucleotide sequence coding for this region of the protein.

The majority of these mutants were constructed by deletion/insertion mutagenesis (Shortle, 1983). A restriction fragment containing sequences coding for the 3′ end of the HA gene was cloned into a plasmid from which all sequences not essential for plasmid viability had been removed. Nicks were introduced randomly into the supercoiled plasmid using DNase I in the presence of ethidium bromide. A short gap was then generated at the site of the single-strand break by incubating the nicked DNA with *Micrococcus luteus* polymerase I. Treatment with single-strand exonuclease from mung bean generated linear molecules to which were ligated *Cla*I synthetic linkers. This procedure yielded closed circular molecules which should have, at the site of the original nick, a small deletion of HA sequence (or essential plasmid sequence) and an insertion of a *Cla*I linker. The mutated plasmids were transfected into *Escherichia coli* strain DH1, and plasmid DNA from the transformants was screened for the presence of a unique *Cla*I restriction site in the sequences coding for the cytoplasmic tail of the HA molecule. Positive clones were sequenced by the chain termination technique (Sanger *et al.*, 1977).

One clone (#164) contained a *Cla*I linker exactly at the junction of the sequences coding for the anchor and tail domains. The deletion/insertion changes the reading frame so that a downstream in-frame termination codon defines an altered tail of 16 novel amino acids. Five other clones contained *Cla*I linkers within the region of interest. However, in all cases, the reading frame was altered such that there were no proximate termination codons and the coding sequences extended far into plasmid or SV40 DNA. To overcome this problem, a 70-base-pair *Taq*I–*Bam*HI DNA fragment from the RSV *env* gene (Schwartz *et al.*, 1983) was inserted between the *Cla*I and *Bam*HI sites of the mutant clones. This fragment codes in one reading frame for the authentic cytoplasmic tail of the env glycoprotein and in others for Glu–Arg–STOP and Lys–Asp–Asp–STOP. It was therefore useful as an adaptor sequence to introduce in-frame terminators after the *Cla*I site. Once the mutants had been constructed and the alterations verified by DNA sequence analysis, the *Eco*RI–*Bam*HI restriction fragment that encompasses the mutated region was used to replace the wild-type DNA fragment in the SV40-HA vector.

In a separate experiment, a final tail mutation was introduced into the HA gene. The restriction enzyme *Nde*I cuts the HA sequence only once within the termination codon of the gene. After digestion of HA DNA with *Nde*I, the protruding 3′ terminus was filled with DNA polmerase I and ligated to pBR322

WILD-TYPE

HA wt ....Ala.Gly.Ile.Ser.Phe.Trp.Met.Cys.Ser.Asn.Gly.Ser.Leu.Gln.Cys.Arg.Ile.Cys.Ile.

PSEUDO WILD-TYPE

HA 71 ....Ala.Gly.Ile.Ser.Phe.Trp.Met.Cys.Ser.Asn.Gly.Ser.Leu.Gln.Cys.SER.Ile.GLU.ARG.

HA 152 env ....Ala.Gly.HIS.ARG.LYS.MET.ILE.ASN.SER.SER.ILE.ASN.TYR.HIS.THR.GLN.TYR.ARG.LYS.MET.GLN.GLY.GLY.ALA.VAL.

SLOWED ER ⟶ GOLGI

HA 11 ....Ala.Gly.Ile.Ser.Phe.Trp.Met.Cys.Ser.ILE.GLU.ARG.

HA 477 env ....Ala.Gly.Ile.Ser.Phe.Trp.Met.Cys.Ser.Asn.Gly.Ser.Leu.Gln.Cys.Arg.Ile.Cys.HIS.ARG.LYS.MET.ILE.ASN.SER...
....SER.ILE.ASN.TYR.HIS.THR.GLN.TYR.ARG.LYS.MET.GLN.GLY.GLY.ALA.VAL.

BLOCKED POST-GOLGI

HA 164 ....Ala.Gly.Ile.Ser.Phe.Trp.Met.Cys.Ser.ILE.ASP.GLY.SER.ALA.GLY.SER.ALA.TYR.ASP.TYR.LYS.SER.PHE.TYR.ASN.

BLOCKED ER

HAxpBR ....Ala.Gly.Ile.Ser.Phe.Trp.Met.Cys.Ser.Asn.Gly.Ser.Leu.Gln.Cys.Arg.Ile.Cys.Ile.SER.PHE.ASN.ALA.VAL.VAL....
....LEU.THR.GLN.SER.GLY.THR.VAL.TYR.GLU.ILE.

FIG. 7. Carboxy-terminal amino acid sequences of HA mutants. Alterations in the cytoplasmic tail are indicated by amino acid codes in all capital letters.

sequences. This generated a mutant HA gene that retains all wild-type sequences and reads through an additional 16 codons until the first in-frame termination codon is encountered in the plasmid sequence. The altered gene was used to replace the wild-type HA sequences in the SV40-HA recombinant.

The C-terminal amino acid sequence encoded by each mutant gene is compared with the wild-type sequence in Fig. 7. Recombinant virus stocks containing the mutant genomes were generated and the mutant proteins were expressed in CV-1 cells. The protein products were quantitated by solid-phase radioimmunoassay and analyzed by immunoprecipitation and SDS–PAGE. All mutant constructs expressed HA in amounts comparable to the wild-type. The HA was detectable only in cell extracts, establishing that none of the altered genes coded for a secreted form of the protein. In several cases an altered mobility of the C-terminal HA2 subunit was detectable on SDS–PAGE, thus confirming the presence of mutant HA. To determine the effect of the mutations on transport, the cellular location of the altered proteins was examined using immunofluorescence (Fig. 8) and erythrocyte binding. In addition, the time course of acquisition by HA of resistance to endo H was determined as a measure of the transit time between the endoplasmic reticulum and the Golgi apparatus. The results are summarized in Table II. In both Fig. 7 and Table II the mutants are ordered with those showing least effect at the top and those causing severe effects at the bottom.

The phenotype of mutant HA-71 is indistinguishable from that of wild-type, displaying normal kinetics of transport. It is perhaps significant that the altered cytoplasmic tail of HA-71 is exactly the same length as that of wild-type HA. The mutation changes three of the four terminal amino acids from Arg–Ile–Cys–Ile to Ser–Ile–Glu–Arg. Thus the high degree of conservation of these amino acids among the HAs from different influenza virus strains (Lamb, 1983) is not a consequence of their importance in intracellular transport. It remains to be deter-

TABLE II
CELLULAR TRANSPORT OF HAs WITH ALTERED CYTOPLASMIC TAILS

	Translocation to ER (signal cleavage) (core glycosylation)	Transport to Golgi: (terminal glycosylation) [Endo H resistance ($t_{\frac{1}{2}}$)]	Transport to surface[a]
HA wild-type	+	+ (20 min)	PM
HA-71	+	+ (20 min)	PM
HA-152env	+	+ (20 min)	PM
HA-11	+	+ (1 hr)	PM
HA-477env	+	+ (1 hr)	PM
HA-164	+	+ (1 hr)	±PM
HA-xpBR	+	− (>8 hr)	−

[a] PM, plasma membrane.

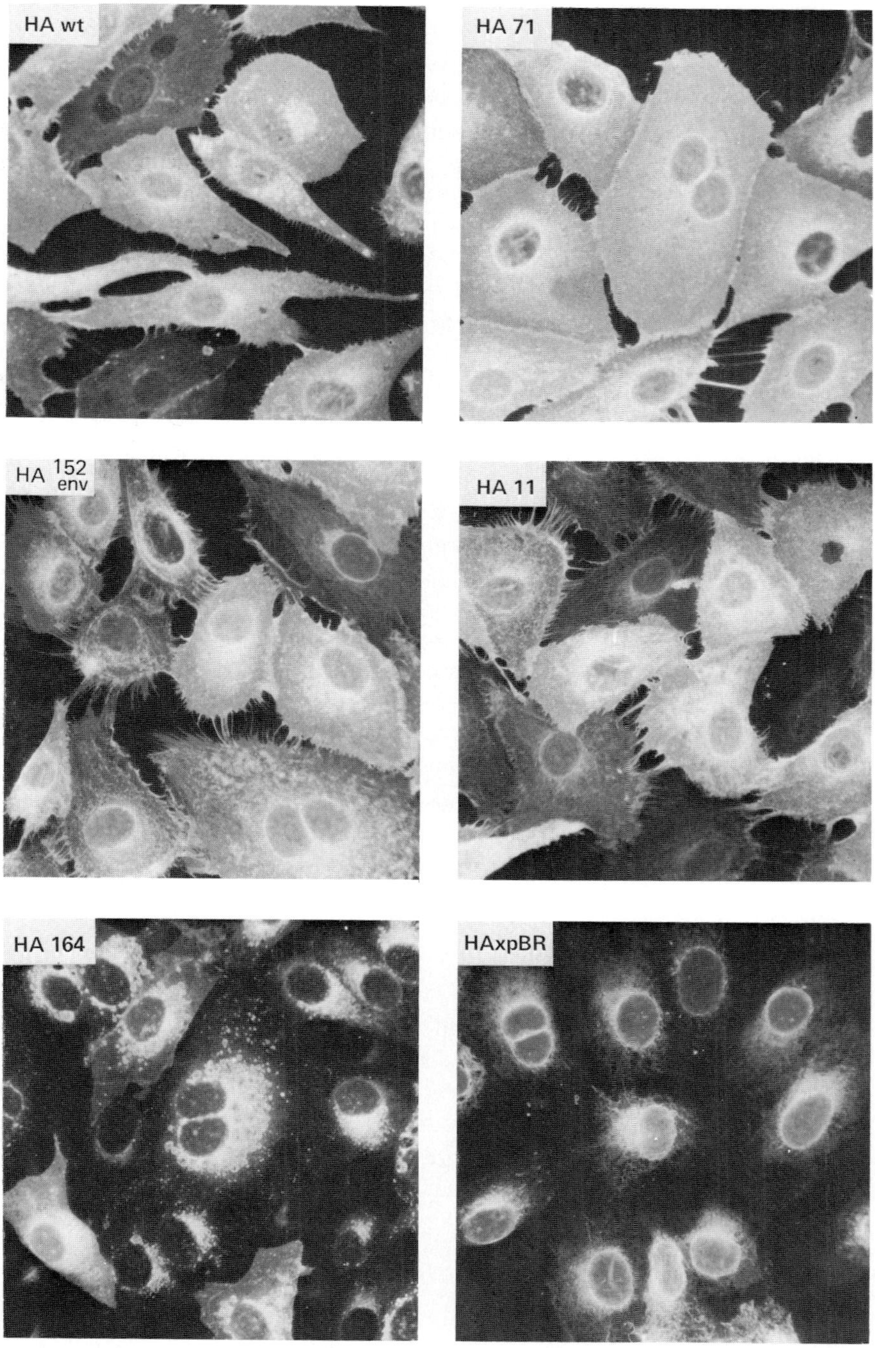

FIG. 8. Immunofluorescent antibody staining on HA in cells infected with recombinant vectors expressing wild-type and mutant HA proteins. Monolayers of CV-1 cells growing on glass slides were infected with the recombinant viruses as described previously (Gething and Sambrook, 1981). Forty hours later the cells were fixed and indirect immunofluorescent staining using rabbit anti-HA serum followed by goat antirabbit IgG conjugated with rhodamine (Cappel) was carried out using the procedure of Ash *et al.* (1977).

mined if these residues are involved in packaging and assembly of HA into the influenza virion. As discussed above, the mutant HA-152env, in which the HA cytoplasmic tail (and seven amino acids of the hydrophobic anchor) has been replaced by the cytoplasmic tail of the RSV env glycoprotein, also has a phenotype indistinguishable from that of wild-type.

A second group of mutants, whose transit between the endoplasmic reticulum and the Golgi is slowed but whose final location on the cell surface is like that of wild-type, includes HA-11 and HA-477env. In HA-11 the 10 amino acid cytoplasmic tail is replaced by a tripeptide Ile–Glu–Arg, while in HA-477env the cytoplasmic tail is intact but extended by the 23 amino acids of the homologous region of the RSV env glycoprotein. A similar phenotype of delayed transport was displayed by several mutants of the VSV glycoprotein which normally has a tail of 29 amino acids. The terminal 16 residues of the tail were replaced by 3, 7 or 23 nonhomologous amino acids (Rose and Bergmann, 1983). Another set of mutants of the Semliki Forest virus (SFV) E2 glycoprotein that have deletions of cytoplasmic tail sequences have been constructed and analyzed by Garoff *et al.* (1983). All the mutant proteins were transported to the cell surface. However, the assays employed would not have distinguished the delayed transit phenotype from normal transport since a small proportion of the HA or VSV mutant proteins arrive at the cell surface as quickly as the wild-types. Thus it is not possible to assign the SFV mutants between the different transport phenotypes.

The transport between the endoplasmic reticulum and the Golgi apparatus of mutant HA-164 was also delayed. However, this mutant also displayed a major defect in transport between the Golgi and the plasma membrane. When erythrocyte binding assays were used to monitor the presence of HA at the surface of infected cells, only approximately 25% of the cells were positive. Indirect immunofluorescence studies using anti-HA serum revealed an unusual pattern of staining (Fig. 8). All the cells contained intensely staining vesicles which were concentrated in an area close to but on one side of the nucleus, and which also radiated less densely throughout the cell. The staining pattern did not coincide with those obtained using antisera directed against endoplasmic reticulum or lysosomes, nor with the fluorescence seen using acridine orange which concentrates in vesicles having a low endogenous pH. However the staining pattern did coincide with that obtained using fluorescent labeled wheat germ agglutinin. Approximately 25% of the cells displayed an additional diffuse cell surface fluorescence. Biochemical analysis using metabolic labeling, immunoprecipitation, and analysis of Endo H sensitivity indicated that all the mutant HA was transported at least to the cis stacks of the Golgi apparatus, since 100% of the protein became endo H resistant within 2 hr. The phenotype of this mutant provides the first indication that a further recognition function may be required to facilitate the transport of the HA molecule from the Golgi apparatus to the cell surface.

The mutant HA-xpBR displayed the most extreme defect. This protein, which has an intact cytoplasmic tail followed by an additional 16 amino acids, appears to be restricted to the endoplasmic reticulum. Indirect immunofluorescence studies using anti-HA serum showed a pattern of staining (Fig. 8) that was identical to that seen using an antiserum directed against the endoplasmic reticulum. No staining was seen in the Golgi or on the cell surface, and cells infected with this mutant did not display erythrocyte binding. Biochemical analysis showed that the protein contained the high-mannose oligosaccharide side chains characteristic of core glycosylation in the endoplasmic reticulum. The glycoprotein remained sensitive to digestion with endo H over a period of 8 hr, indicating that it was not transported to the Golgi apparatus. A comparable mutant of the VSV G protein, which contains a cytoplasmic tail extended to 38 amino acids from its normal length of 29 residues, has a very similar phenotype (Rose and Bergmann, 1983). In addition, two further VSV G mutants, in which the entire cytoplasmic tail has been replaced by either 3 or 12 nonhomologous amino acids, are not transported beyond the endoplasmic reticulum.

The receptor model described earlier would predict that the various mutants that are delayed in transport to the Golgi have decreased affinities for receptor(s) responsible for transporting the nascent glycoproteins from the endoplasmic reticulum to the Golgi. If this is so, the question remains as to whether these different viral glycoproteins (and other glycoproteins normally synthesized in CV-1 cells) interact with the same or distinct receptor molecules. The lack of homology between the primary sequences of the proteins argues for different receptors, although it is possible that the molecules may fold to form a structural feature common to all proteins destined to leave the endoplasmic reticulum. At this stage it is not possible to conclude whether the cytoplasmic tails themselves might contain the recognition site(s) for such receptor(s) or whether alterations in the tail cause conformational changes that perturb recognition in the other protein domains.

In the absence of any definitive evidence for interaction of HA with receptor or carrier proteins, the simplest explanation of the phenotypes of the mutants is that alterations in the primary sequence of the cytoplasmic tail can act directly to retard the movement of the molecule. For example, deletion or extension of the cytoplasmic tail might destabilize interactions between the three transmembrane domains of the HA trimer, which are thought to cross the lipid bilayer as stacked α helices. This might prevent or delay folding and oligomerization of the entire molecule—presumably a prerequisite for its movement. Alternatively, the disordered transmembrane structure could retard the movement of the protein in the plane of the membrane or obstruct the incorporation of the protein into transport vesicles that shuttle it from the endoplasmic reticulum to (and through) the Golgi apparatus. Totally novel sequences such as those appended to the tail of *HAxpBR* or *HA164* might be expected to cause more disruption than sequences that have

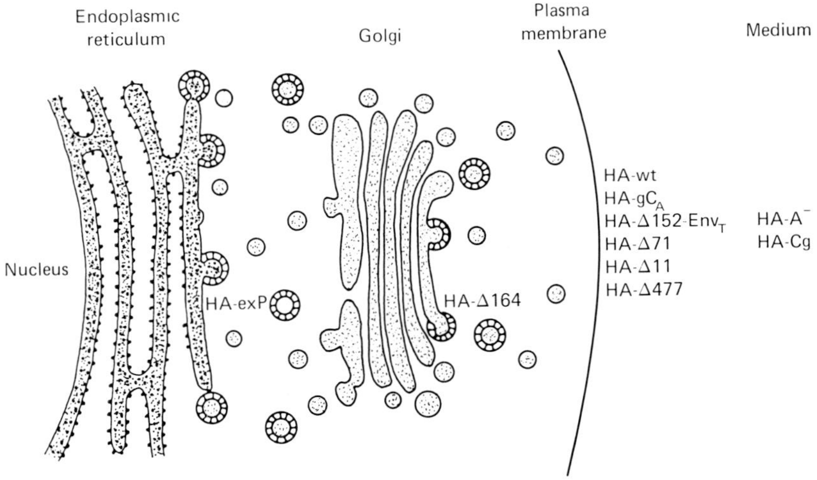

FIG. 9. Diagrammatic representation of the cellular organelles involved in the transport of nascent glycoproteins to the cell surface. The final locations of the wild-type and mutant HA molecules are shown on the diagram.

evolved as cytoplasmic tails on other membrane proteins. According to this hypothesis, substitution or addition of the env cytoplasmic sequences might be expected to have comparatively little effect on the rate of transport of HA.

E. Summary

Figure 9 summarizes the results that we have obtained using *in vitro* mutagenesis of cloned HA genes to study the function of various domains of the HA molecule in determining its transport to the cell surface. Although these studies are far from complete, the work testifies to the power of this technology, which can be extended to study the role of other portions of the molecule, including the fusion peptide, the antigenic sites, the receptor binding site, and the carbohydrate attachment points. Furthermore, precise dissection of the contribution of individual amino acids to the structure and function of the molecule can be achieved by introducing single nucleotide changes at defined sites within the HA gene using oligonucleotide-directed mutagenesis (Zoller and Smith, 1983). Although these experiments will be of particular relevance to our analysis of the HA molecule, the results will in addition be of great value in our understanding of the structure and function of integral membrane proteins in general.

REFERENCES

Air, G. M. (1981). Sequence relationships among the hemagglutinin genes of 12 subtypes of influenza A virus. *Proc. Natl. Acad. Sci. U.S.A.* **78,** 7639–7643.

Ash, J. F., Louvard, D., and Singer, S. J. (1977). Antibody-induced linkages of plasma membrane proteins to intracellular actomyosin-containing filiaments in cultured fibroblasts. *Proc. Natl. Acad. Sci. U.S.A.* **74,** 5584–5588.

Blobel, G., Walter, P., Chang, C. N., Goldman, B. M., Erikson, A. H., and Lingappa, V. R. (1979). Translocation of proteins across membranes: the signal hypothesis and beyond. *Symp. Soc. Exp. Biol.* **33,** 9–36.

Davis, B. D., and Tai, P.-C. (1980). The mechanism of protein secretion across membranes. *Nature (London)* **283,** 433–438.

Fitting, T., and Kabat, D. (1982). Evidence for a glycoprotein "signal" involved in transport between subcellular organelles. *J. Biol. Chem.* **257,** 14011–14017.

Frink, R. J., Eisenberg, R., Cohen, J., and Wagner, E. K. (1983). Detailed analysis of the portion of the Herpes Simplex virus type 1 genome encoding glycoprotein C. *J. Virol.* **45,** 634–647.

Garoff, H., Kondor-Koch, C., Pettersson, R., and Burke, B. (1983). Expression of Semliki Forest virus proteins from cloned complementary DNA. II. The membrane spanning glycoprotein E2 is transported to the cell surface without its normal cytoplasmic domain. *J. Cell Biol.* **97,** 652–658.

Garten, W., Bosch, F.-X., Linder, D., Rott, R., and Klenk, H.-D. (1981). Proteolytic activation of the influenza virus hemagglutinin: the structure of the cleavage site and the enzymes involved in cleavage. *Virology* **115,** 361–374.

Gething, M. J., and Sambrook, J. (1981). Cell surface expression of influenza haemagglutinin from a cloned DNA copy of the RNA gene. *Nature (London)* **293,** 620–625.

Gething, M. J., and Sambrook, J. (1982). Construction of influenza haemagglutinin genes that code for intracellular and secreted forms of the protein. *Nature (London)* **300,** 598–603.

Gething, M. J., and Sambrook, J. (1983). Expression of cloned influenza virus genes. *In* "Genetics of Influenza Viruses" (P. Palese and D. W. Kingsbury, eds.), pp. 169–191. Springer-Verlag, Berlin and New York.

Gething, M. J., White, J. M., and Waterfield, M. D. (1978). Purification of the Fusion protein of Sendai virus: Analysis of the NH_2-terminal sequence generated during precursor activation. *Proc. Natl. Acad. Sci. U.S.A.* **75,** 2737–2740.

Gething, M. J., Bye, J., Skehel, J. J., and Waterfield, M. D. (1980). Cloning and DNA sequence of double-stranded copies of haemagglutinin genes from H2 and H3 strains elucidates antigenic shift and drift in human influenza virus. *Nature (London)* **287,** 301–306.

Gluzman, Y. (1981). SV40 transformed simian cells support the replication of early SV40 mutants. *Cell* **23,** 175–182.

Hartman, J. R., Nayak, D. P., and Fareed, G. C. (1982). Human influenza virus Haemagglutinin is expressed in monkey cells using simian virus 40 vectors. *Proc. Natl. Acad. Sci. U.S.A.* **79,** 233–237.

Heiland, I., and Gething, M. J. (1981). Cloned copy of the haemagglutinin gene codes for human influenza antigenic determinants in *E. coli*. *Nature (London)* **292,** 851–852.

Hirst, G. K. (1941). Agglutination of red cells by allantoic fluid of chick embryos infected with influenza virus. *Science* **94,** 22–23.

Hubbard, S. C., and Robbins, P. W. (1979). Synthesis and processing of protein linked oligosaccharides *in vivo*. *J. Biol. Chem.* **254,** 4568–4576.

Lamb, R. A. (1983). The influenza virus RNA segments and their encoded proteins. *In* "Genetics of Influenza Viruses" (P. Palese and D. W. Kingsbury, eds.), pp. 21–69. Springer-Verlag, Berlin and New York.

Lodish, F. H., Kong, N., Snider, M., and Strous, G. J. A. M. (1983). Hepatoma secretory proteins migrate from the rough endoplasmic reticulum to the Golgi at characteristic rates. *Nature (London)* **304,** 80–83.

McCutchan, J. H., and Pagano, J. S. (1968). Enhancement of the infectivity of simian virus 40 deoxyribonucleic acid with diethyl aminoethyl-dextran. *J. Natl. Cancer Inst.* **41,** 351–357.

Matlin, K., Reggio, H., Helenius, A., and Simons, K. (1981). The infective entry of influenza virus into MDCK cells. *J. Cell Biol.* **91,** 601–613.

Neuberger, A., Gottschalk, A., Marshall, R. O., and Spiro, R. G. (1972). "The Glycoproteins: Their Composition, Structure and Functions," pp. 450–490. Elsevier, Amsterdam.

Pipas, J. M., Adler, S. P., Peden, K. W. C., and Nathans, D. (1979). Deletion mutants of SV40 that affect the structure of viral tumor antigens. *Cold Spring Harbor Symp. Quant. Biol.* **44,** 285–291.

Richardson, C. D., Scheid, A., and Choppin, P. W. (1980). Specific inhibition of paramyxovirus and myxovirus replication with oligopeptides with amino acid sequences similar to those at the N-termini of the F_1 or HA_2 viral polypeptides. *Virology* **105,** 205–222.

Rose, J. K., and Bergmann, J. E. (1982). Expression from cloned cDNA of cell-surface and secreted forms of the glycoprotein of vesicular stomatitis virus in eucaryotic cells. *Cell* **30,** 753–762.

Rose, J. K., and Bergmann, J. E. (1983). Altered cytoplasmic domains affect intracellular transport of the vesicular stomatitis virus glycoprotein. *Cell* **34,** 513–524.

Roth, M. G., Compans, R. W., Giusti, L., Davis, A. R., Nayak, D. P., Gething, M. J., and Sambrook, J. (1983). Influenza virus hemagglutinin expression is polarized in cells infected with recombinant SV40 viruses carrying cloned hemagglutinin DNA. *Cell* **33,** 435–443.

Sabatinin, D. D., Kreibich, G., Morimoto, T., and Adesnik, M. (1982). Mechanisms for th incorporation of proteins in membranes and organelles. *J. Cell Biol.* **92,** 1–22.

Sanger, F., Nicklen, S., and Coulson, A. R. (1977). DNA sequencing with chain terminating inhibitors. *Proc. Natl. Acad. Sci. U.S.A.* **74,** 5463–5471.

Schlesinger, M. J. (1981). Proteolipids. *Annu. Rev. Biochem.* **50,** 193–206.

Schwartz, D. E., Tizard, R., and Gilbert, W. (1983). Nucleotide sequence of Rous sarcoma virus. *Cell* **32,** 853–858.

Shortle, D. (1983). A genetic system for analysis of Staphlococcal nuclease. *Gene* **22,** 181–189.

Skehel, J. J., and Waterfield, M. D. (1975). Studies on the primary structure of the influenza virus haemagglutinin. *Proc. Natl. Acad. Sci. U.S.A.* **72,** 93–97.

Strous, G. J. A. M., and Lodish, H. F. (1980). Intracellular transport of secretory and membrane proteins in hepatoma cells infected by vesicular stomatitis virus. *Cell* **22,** 709–717.

Strous, G. J. A. M., Willemsen, R., van Kerkoff, P., Slot, J. W., Geuze, H. J., and Lodish, H. F. (1983). Vesicular stomatitis virus glycoprotein, albumin, and transferrin are transported to the cell surface via the same Golgi vesicles. *J. Cell Biol.* **97,** 1815–1822.

Sveda, M. M., and Lai, C. J. (1981). Functional expression in primate cells of cloned DNA coding for the halmagglutinin surface glycoprotein of influenza virus. *Proc. Natl. Acad. Sci. U.S.A.* **78,** 5488–5492.

Sveda, M. M., Markoff, L. J., and Lai, C. J. (1982). Cell surface expression of the influenza virus haemagglutinin requires the hydrophobic carboxy-terminal sequences. *Cell* **30,** 649–656.

Tabas, I., and Kornfeld, I. J. (1979). Purification and characterization of a rat liver golgi β-mannosidase capable of processing aparagine-linked oligosaccharides. *J. Biol. Chem.* **254,** 11655–11663.

Tarentino, A. L., and Maley, F. (1974). Purification and properties of an endo-β-glucosaminidase from *Streptomyces griseus*. *J. Biol. Chem.* **249,** 811–817.

Tooze, J., ed. (1980). "DNA Tumor Viruses," 2nd Ed. Cold Spring Harbor Laboratory, Cold Spring Harbor, New York.

Walter, P., and Blobel, G. (1981). Translocation of proteins across the endoplasmic reticulum II. Signal recognition protein (SRP) mediates the selective binding to microsomal membranes of *in vitro* assembled polysomes synthesizing secretory proteins. *J. Cell Biol.* **91,** 551–558.

White, J. M., Helenius, A. A., and Gething, M. J. (1982). The haemagglutinin of influenza virus expressed from a cloned gene promotes membrane fusion. *Nature (London)* **300,** 658–659.

Wiley, D. C., Skehel, J. J., and Waterfield, M. D. (1977). Evidence from studies with a cross-linking reagent that the haemagglutinin of influenza virus is a trimer. *Virology* **79,** 446–448.

Wiley, D. C., Wilson, I. A., and Skehel, J. J. (1981). Structural identification of the antibody binding sites of the Hong Kong influenza haemagglutinin and their involvement in antigenic variation. *Nature (London)* **289,** 373–378.

Wilson, I. A., Skehel, J. J., and Wiley, D. C. (1981). The haemagglutinin membrane glycoprotein of influenza virus: structure at 3 Å resolution. *Nature (London)* **289,** 366–373.

Zoller, M. J., and Smith, M. (1983). Oligonucleotide-directed mutagenesis of DNA fragments cloned into M13 derived vectors. *In* "Methods in Enzymology" (R. Wu, L. Grossman, and K. Moldave, eds.), Vol. 100, pp. 468–500. Academic Press, New York.

Part II

Channels

CURRENT TOPICS IN MEMBRANES AND TRANSPORT, VOLUME 23

Chapter 3

Ca^{2+} Channels of *Paramecium:* A Multidisciplinary Study

*CHING KUNG**,† AND YOSHIRO SAIMI**

**Laboratory of Molecular Biology and* †*Department of Genetics*
University of Wisconsin
Madison, Wisconsin

Like higher animals, the eukaryotic unicell, *Paramecium,* responds actively to touch, heat, chemicals, and light (Jennings, 1906). The central mechanism of these responses is an action potential (Kinosita *et al.*, 1964). It is therefore likely that the machinery for signal transduction in the major sensory modalities, for membrane excitation, for excitation–motility coupling, and even that for adaptation had already evolved prior to the divergence of proto- and metazoa. Although neurophysiological research has centered on cell–cell communication in the neural network, interest in protozoan behavior is more than esoteric because its underlying biochemical and biophysical principles are probably the same as in metazoan behavior. We have forfeited the option of studying cell–cell communication for the advantages in experimentation toward an understanding of these underlying principles. The advantages of working with *Paramecium* are basically two (Kung, 1971a). First, it is a large cell: *Paramecum caudatum* is 200 μm in length; *Paramecium tetraurelia* 100 μm. This cell size allows intracellular re-

ISBN 0-12-153323-9

cording with low-resistance electrodes (Naitoh and Eckert, 1972) and routine microinjection of foreign material into the cytoplasm (Haga *et al.*, 1982). Second, it grows rapidly and to large number, doubling every 6–8 hr and yielding grams of cells of the same genotype and phenotype. Consequently, genetic (Sonneborn, 1970) and biochemical studies (Nelson and Kung, 1978) are relatively easy. This article covers multidisciplinary research on the ion channels, especially the Ca^{2+} channel, of *Paramecium*. For other articles with different emphases, see Adoutte *et al.* (1981), Byrne and Byrne (1978), Cronkite (1979), Doughty and Dryl (1981), Eckert and Brehm (1979), Kung (1979), Kung and Saimi (1982), Nelson and Kung (1978), and Van Houton *et al.* (1981).

I. A SWIMMING NEURON

Each paramecium is completely covered by thousands of cilia with which it swims about. Each cilium is made of an axoneme (the 9 + 2 microtubular assembly and associated structures) and its membrane cover. The axoneme alone, without the membrane, can beat as long as Mg^{2+} and ATP are provided (Naitoh and Kaneko, 1972). Force is generated by the dynein arms, which have Mg^{2+}-ATPase activity and are attached to the peripheral microtubules. How the tens of thousands of dynein arms along and around each axoneme are regulated to generate the rhythmic beat remains unclear (Gibbons, 1981; Omoto and Kung, 1979). The form, direction, and frequency of the beat can be changed leading to sprinting, stoppage, or backward swimming of the paramecium, i.e., its behaviors.

The ciliary membrane is continuous with the body surface membrane. Together they contain the receptors and the ion channels. Certain cations, organic repellents, mechanical stimulation at the anterior end, or heat cause a depolarization, which, in turn, triggers a Ca^{2+} action potential. The Ca^{2+} entry increases the internal free Ca^{2+} concentration. Upon the elevation of the Ca^{2+} concentration, the axonemes beat more rapidly and at an angle up to 100° from the usual direction (ciliary reversal, Fig. 1). The system returns to normal after the action potential subsides and the Ca^{2+} removed, presumably by a Ca^{2+} pump. The observed behavior is the interruption of the left-handed helical forward progress by a period of right-handed helical backward swimming. This "avoiding reaction" can be of different strengths, which are graded to the strengths of the stimuli. This sequence of events was analyzed in detail in the 1960s and 1970s by Naitoh, Eckert, and co-workers (Eckert *et al.*, 1976).

A parallel pathway is also known. Mechanical stimulation at the posterior end, certain ions, or certain attractants cause a K^+-based hyperpolarization. This negative shift in membrane causes only a slight change in ciliary beat direction

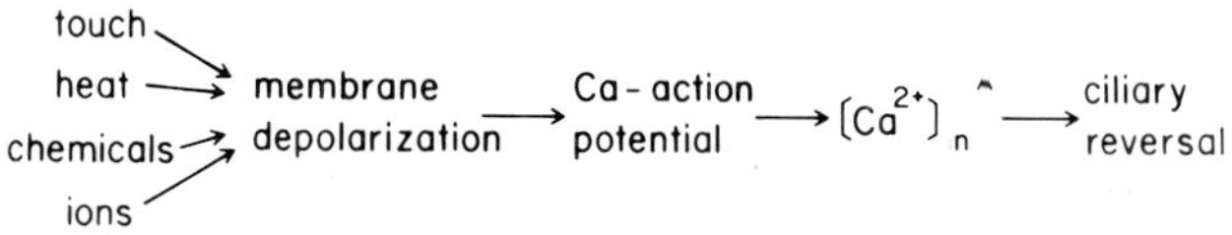

FIG. 1. A sensory transduction pathway of *Paramecium*.

but increases the beat frequency significantly. The result is a forward sprint of the paramecium, the "escape reaction" (Naitoh and Eckert 1969, 1973; Naitoh, 1974). The messenger mediating the hyperpolarization and the axoneme has not been identified although recent studies implicate cAMP (Gustin *et al.*, 1983).

A paramecium can therefore be viewed as a one-called conglomerate of a sense cell, a neuron, and an effector cell, although the evolutionary process is likely to be a diversification of a mother corporation instead of a merger. This "swimming sensory cell," as Machemer and de Peyer (1977) dubbed it, exhibits its membrane electric states as overt behaviors. Manipulation and measurement of these behaviors approximate the manipulation and measurement of electrophysiological parameters of as many as 10^7 cells conveniently, simultaneously, and inexpensively.

II. ION CHANNELS

The form of the action potential in *Paramecium* varies with the stimuli and composition of the external solution. In a conventional bath of 4 m*M* KCl, 1 m*M* $CaCl_2$, 1 m*M* HEPES, pH 7.0, the membrane rests at −40 mV inside negative. Depolarization by injected outward current (or by other means) triggers an action potential graded to the strength of the stimulus. The maximal amplitude is about 50 mV. The upstroke of this action potential is due to Ca^{2+} influx and the downstroke due to K^+ efflux (Naitoh and Eckert, 1969) (Fig. 2A). External Na^+ makes the membrane more excitable and encourages spontaneous discharges (Satow and Kung, 1974).

Although the active membrane depolarizations and hyperpolarizations are the essential parts of the behavior, the activities of the ion channels are better characterized by their currents when the voltage is experimentally controlled. Under a voltage clamp, a step depolarization from −40 mV induces a transient Ca^{2+} current (Fig. 2C) (Oertel *et al.*, 1977; Satow and Kung, 1979; Brehm *et al.*, 1980). A step to −10 mV induces the maximal Ca^{2+} transient of about 7 nA in *P. tetraurelia* or 10 nA in *P. caudatum*. The Ca^{2+} current decays exponentially during the depolarization with a time constant of 1 msec toward a level of less than 1 nA. This decay is due to the internal Ca^{2+}, which comes through the channel itself, and is termed the Ca^{2+}-dependent Ca^{2+}-channel inactivation

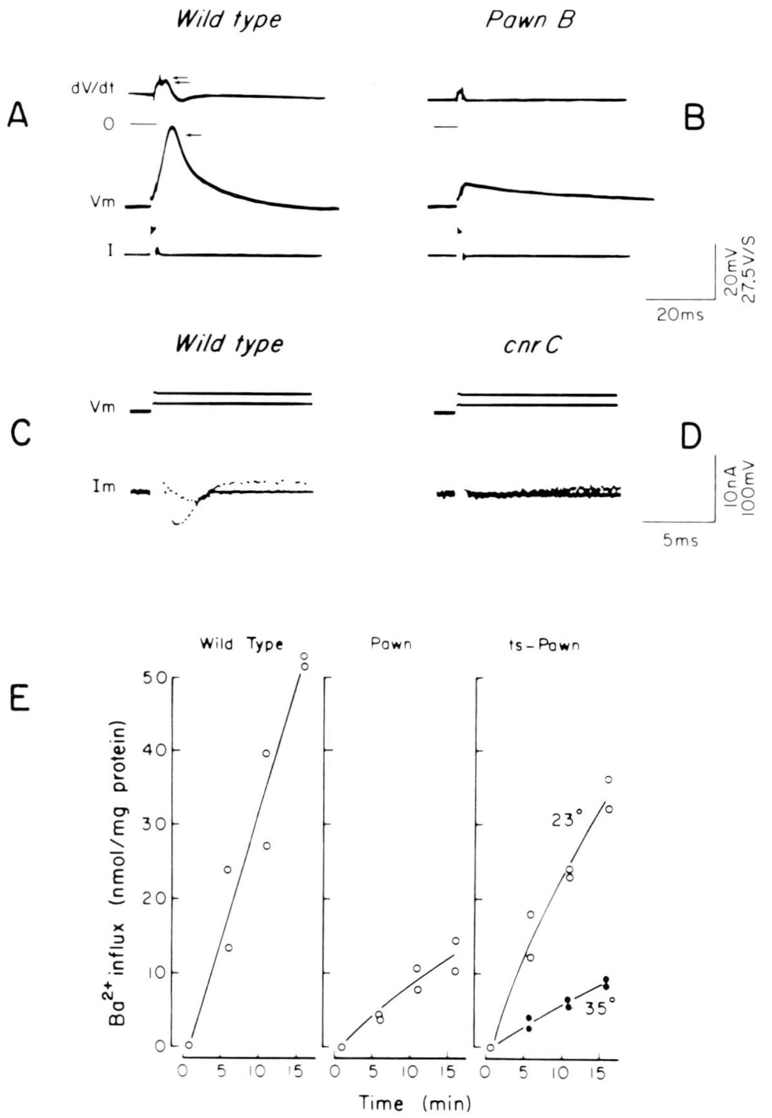

FIG. 2. Membrane-inexcitable mutants of *Paramecium*. (A) A brief outward current, *I*, injected across the membrane of a wild-type *P. tetraurelia* induces an action potential (arrow, *V*m trace, 0 is the reference level) with an active component seen in the *dV*/*dt* trace (double arrow). (B) Same treatment of a *pawnB* mutant yields only passive responses. Cells bathed in 4 m*M* K^+, 1 m*M* Ca^{2+}, 1 m*M* Tris, pH 7.2. (see Kung, 1976). (C) Under a voltage clamp a step depolarization from −44 to −30 or −15 mV (*V*m traces) induces a small or a large Ca^{2+}-inward current (*I*m traces) in the wild-type *P. caudatum*. (D) The inexcitable mutant *cnrC* does not generate the Ca^{2+} current upon step

(Brehm and Eckert, 1978a) and may require binding of Ca^{2+} to the channel. Barium ions, which also pass through the channel, cannot inactivate it. Not only this inactivation but also activation (the rise of current upon step depolarization) and deactivation (the tail current upon step repolarization) are also affected by the permeant ions since the kinetics of these processes vary depending on whether Ca^{2+}, Sr^{2+}, or Ba^{2+} carries the current through the Ca^{2+} channel (Saimi and Kung, 1982). Like those of other cells, the Ca^{2+} current of *Paramecium* can be reduced by La^{3+} or Cd^{2+}, but verapamil, D-600, or nifedipine has little effect (Y. Saimi, unpublished results). The Ca^{2+} current of *Paramecium* is inhibited by W-7, a compound that inhibits the activity of calmodulin and related structures (Hennessey and Kung, 1984). The voltage-dependent Ca^{2+} channels are located exclusively in the ciliary membrane since deciliation deletes the action potential and the action current while reciliation restores them (Ogura and Takahashi, 1976; Dunlap, 1977; Machemer and Ogura, 1979).

There are several other ion channels on the surface membrane of *Paramecium*. They are defined by their currents, which have different ion specificities, triggering mechanisms, and activation and inactivation kinetics. Among these are the depolarization-dependent K^+ current (the delayed rectifier) and the hyperpolarization-dependent K^+ current (the anomalous rectifier) (Naitoh and Eckert, 1968; Satow and Kung, 1980a; Oertel *et al.*, 1978). There are also two slower currents that are activated by Ca^{2+} instead of voltage: the Ca^{2+}-dependent K^+ current (Satow, 1978; Satow and Kung, 1980a) and the Ca^{2+}-dependent Na^+ current (Saimi and Kung, 1980). The sensory receptors have separate channels. There is a somal divalent-ion current activated by anterior mechanical stimulation and a K^+ current activated by posterior stimulation (Naitoh and Eckert, 1969, 1973; Eckert *et al.*, 1972; Ogura and Machemer, 1980; Satow *et al.*, 1983). Reception of organic attractants or repellents is followed by voltage changes (Van Houten, 1979, 1981). Heat also induces a depolarization (Hennessey *et al.*, 1983), although it is not clear whether there is a special receptor. The various ion currents in *Paramecium* have recently been reviewed (Table 1 of Kung and Saimi, 1982). They have been characterized by conventional electrophysiological methods, by channel blockers, and by using mutations to eliminate or enhance various currents.

depolarizations. Cells bathed in 1 m*M* K^+, 1 m*M* Ca^{2+}, 10 m*M* TEA^+, 1 m*M* HEPES, and 10^{-2} m*M* EDTA, pH 7.3 (see Haga *et al.*, 1983). (E) $^{133}Ba^{2+}$ accumulation in wild-type *P. tetraurelia* (left) over time when washed cells were incubated with a Ba^{2+}-containing solution. The rate and extent of that uptake are smaller in the *pawnB* mutant (center). The *pawnC* mutant with a temperature-sensitive mutation has an intermediate uptake when grown at a permissive temperature (23°C) but is defective when grown at a restrictive temperature (35°C) All tests were performed at room temperature, 23°C (see Ling and Kung, 1980, for details).

III. MUTANTS

As in contemporary investigations of many other biological systems, mutational defects are induced in the behavior of *Paramecium* to delineate mechanistic pathways, block individual steps, and tag the gene products needed in those steps (Kung 1971a,b; Takahashi, 1979). *Paramecium* is somatically polyploid but genetically diploid (Sonneborn, 1975). Unlike most diploid metazoa, it can self-fertilize readily during the induced selfing-conjugation of *P. caudatum* (Hiwatashi, 1969) and the naturally occurring autogamy in *P. tetraurelia* (Sonneborn, 1970). Self-fertilization in diploid organisms greatly enhances the probability of expressing and thereby recovering recessive mutations. Autogamy, the extreme self-fertilization of two identical mitotic daughters of the same haploid nucleus, raises that probability to 0.5. The advantage in recovering mutations has been fully exploited to generate mutants defective in behavior. After treatment with a mutagen (nitrosoguanidine, X rays, or gamma rays) and the induction of self-fertilization, populations of paramecia are screened for mutants. They are selected by behavioral methods [picking misbehaving variants in geotactic columns or galvanotactic troughs filled with appropriate ion solutions (Kung, 1971a; Takahashi, 1979; Hinrichsen *et al.*, 1984b)], by survival methods [rescuring survivors after incubation in high doses of ions or drugs that are lethal to the wild type (Schein, 1976a; Shusterman *et al.*, 1978)], or by chance.

There are *Paramecium* mutants defective in sensory processes. Van Houten (1977) and DiNallo *et al.* (1982) have described mutants defective in acetate or folate chemotaxis but normal otherwise. There are also mutants defective in their motility. The more interesting ones of this group are those whose cilia, though fully motile, fail to respond properly to Ca^{2+} although all steps up to the delivery of Ca^{2+} (Fig. 1) are normal (Hinrichsen and Kung, 1984; Hinrichsen *et al.*, 1984a).

Of special interest to the topic covered by this volume are mutants of a third type. They are those with abnormal membrane currents indicating channel malfunction. Mutants defective in their Ca^{2+} current are reviewed in the following sections in greater detail. There are, however, many mutants that are defective in other ion currents. Among them are the "pantophobiac" mutants (three isolates: *pant*), which have lost most of their Ca^{2+}-dependent K^+ current (Saimi *et al.*, 1983); the "paranoiac" mutants (seven loci: *PaA, paB, PaC, PaD, PaE, PaF, fnaP*) (Van Houten *et al.*, 1977), one of which (*PaA*) has been shown to have a stronger Ca^{2+}-dependent Na^+ current and a larger $^{22}Na^+$ influx (Saimi and Kung, 1980; Hansma and Kung, 1976; Satow *et al.*, 1976); *baA* (three alleles), which is defective in all the voltage-dependent channels and has a change in a set of sphingolipids and phosphonolipids (Forte *et al.*, 1981); *teaB* (one allele), with a 10-mV shift of I–V relations of all voltage-dependent currents (Satow and Kung, 1981); *fna* (5 alleles), which tends to hyperpolarize in certain solutions

(Satow and Kung, 1976c); and *teaA* (one allele), with a stronger rectifying K^+ current (Satow and Kung, 1976b).

IV. *PAWNS* AND *CNR*s

These are mutants with little or no Ca^{2+} current. They are therefore without action potentials and cannot increase their internal Ca^{2+} concentration in response to depolarizing stimuli. They cannot swim backward in the face of these stimuli, hence the names *pawn* (after the chess piece) for the *P. tetraurelia* mutants and *CNR* (caudatum nonreversal) for the *P. caudatum* mutants (Kung, 1971a,b; Takahashi, 1979).

Outward currents injected through the membranes of these mutants produce only the passive response (Fig. 2B) but not the action potential seen in the wild type (Fig. 2A) (Kung and Eckert, 1972; Schein *et al.*, 1976; Takahashi and Naitoh, 1978). Step depolarizations with a voltage clamp produce leakage current and activate the outward K^+ current (Fig. 2D) in these mutants but not the inward Ca^{2+} current of the wild-type (Fig. 2C) (Oertel *et al.*, 1977; Satow and Kung, 1980b; Takahashi and Naitoh, 1978). The loss of Ca^{2+}-channel function in these mutants has also been demonstrated by the failure to accumulate $^{45}Ca^{2+}$ (Browning and Nelson, 1976; Browning *et al.*, 1976) or $^{133}Ba^{2+}$ properly (Ling and Kung, 1980) (Fig. 2E). Other voltage-dependent currents which include the delayed rectifier and the anomalous rectifier remain intact except in the case of *pawn B* (below) (Kung and Eckert, 1972; Schein *et al.*, 1976; Takahashi and Naitoh, 1978). The mutational defects also do not extend to the axoneme. Detergent-treated models of these mutants remain capable of swimming backward when Ca^{2+} is added (Kung and Naitoh, 1973; Takahashi and Naitoh, 1978), reinforcing the notion that the failure to respond to stimuli is due not to the failure of the axonemes to respond to Ca^{2+} but to the failure of the membrane to let in Ca^{2+}. Besides the ciliary response, the Ca^{2+}-activated currents are also indicators of an increase in internal Ca^{2+} concentration. As expected, both the Ca^{2+}-dependent K^+ current (Satow and Kung, 1980a) and the Ca^{2+}-dependent Na^+ current (Saimi and Kung, 1980) are greatly curtailed indirectly by these mutations which cut down the Ca^{2+} current.

Genetically the *pawns* of *P. tetraurelia* belong to three complementation groups: *pwA*, *pwB*, and *pwC* (Kung 1971b; Chang *et al.*, 1974; Schein, 1976a). Over 200 lines of *pwA* and *pwB* have been isolated together with several *pwC*s. The mutant genes are found to be very recessive to their wild-type alleles based on simple behavioral tests of the heterozygotes. The three genes are unlinked. The *pwA* mutants range from severe, leaky to conditional (Chang *et al.*, 1974; Schein *et al.*, 1976). The conditional mutants are dependent on growth temperatures. They are quasi-normal when grown at the permissive temperatures

(Chang and Kung, 1973a,b; Satow *et al.*, 1974; Satow and Kung, 1976a, 1980b) but lose the Ca^{2+} channel function completely when grown at restrictive temperatures (Fig. 2E) (Browning *et al.*, 1976; Ling and Kung, 1980). The *pwC* mutants are all temperature sensitive in a similar way. The *pwB* mutants generally tend to be nonleaky and no temperature-sensitive alleles have yet been discovered. The *pwB* mutants, but not *pwA* or *pwC*, are also defective in the anomalous rectification (Schein *et al.*, 1976) and survive concentration of K^+ lethal to the wild type (Shusterman *et al.*, 1978). The bases of these pleiotropic effects of *pwB* are not known.

The *CNR* mutants of *P. caudatum* are also genetically separable into three complementation groups: *cnrA*, *cnrB*, and *cnrC* (Takahashi, 1979). They are also recessive and unlinked to one another. *cnrB* has an allele $cnrB^{ks}$ (K^+ sensitive), which, instead of failing to swim backward in the depolarizing K^+ solution, swims backward for much longer periods than does the wild type. This is interesting because it suggests two opposite changes in the number or the character of the Ca^{2+} channels by two alleles of the same gene. The genetic relation of the three *pw* genes and the three *cnr* genes cannot be tested by breeding since interspecific conjugation, although artificially inducible, is infertile. Tests for functional complementation through transfusion by microinjection (below) show, surprisingly, that all six genes are independent of one another (Haga *et al.*, 1983).

Pawns and *Cnr*s provide the null control in various experiments in which the flow of Ca^{2+} or its consequences in *Paramecium* are to be analyzed. They were used in the separation of the Ca^{2+} current from the K^+ currents (Oertel *et al.*, 1977), in the separation of the Ca^{2+}-dependent K^+ current from other currents (Satow and Kung, 1980a), in isolating the depolarizing mechanoreceptor potential (Satow *et al.*, 1983), in monitoring the flow of Ca^{2+} or Ba^{2+} through the Ca^{2+} channel into whole cells *in vivo* (Browning and Nelson, 1976; Browning *et al.*, 1976; Ling and Kung, 1980) and into ciliary-membrane vesicles *in vitro* (Thiele and Schulz, 1981; Thiele *et al.*, 1983), in relating hydrostatic pressure to channel function (Otter and Salmon, 1979), in assessing the need for the avoiding reaction in chemotaxis (Van Houten *et al.*, 1975, 1981) and thermotaxis (Hennessey and Nelson, 1979), in investigating the coupling between frequency control and direction control of ciliary beat (Brehm and Eckert, 1978b), and in assessing the lifetime of Ca^{2+} channel in the membrane (Schein, 1976b).

The lack of Ca^{2+} current in the severe mutants prevents further analysis of the defects in the current. Examination of the residual currents in the leaky or temperature-sensitive mutants of *pwA* showed no gross changes in the voltage at which maximal current is seen nor in their Ba^{2+} current characters, suggesting no gross change in voltage sensitivity or ion selectivity by the *pwA* mutations (Satow and Kung, 1980b). At present, we do not know whether the *pw* and the *cnr* genes code for the Ca^{2+} channel structure, code for elements needed for proper functions, or control the synthesis or assembly of the channels.

V. CURING FACTORS

Although the ciliary membrane can be purified (see below), direct electrophoretic comparison of the membrane proteins of wild type and of mutants in a search for the pawn gene products gave inconclusive results (Merkel *et al.*, 1981; Adoutte *et al.*, 1981, 1983).

A more profitable approach when seeking the gene products is to look in the wild type for factors that restore the Ca^{2+} channel function in the *pw* or *cnr* mutants. Berger (1976) observed that, during a conjugation of wild type and *pwA*, the wild mate confers the backward-swimming ability to the *pwA* partner even before cross-fertilization and long before the expression of the newly received genes. This observation suggests that the wild-type conjugant provides a diffusible substance that restores the Ca^{2+} channel function in the *pwA* mutant. Hiwatashi and co-workers (1980) took this observation one step further: they took the cytoplasm directly from wild-type paramecia, injected it into that of the *cnrC* mutant cells, and observed the restoration of the backward-swimming capability. This observation was later extended to all the inexcitable mutants of both species (Haga *et al.*, 1982). Typically, about 10% of the cell volume (about 50 pl) is transferred. The ability to swim backward becomes measurable less than 1 hr after injection. Near wild-type capability is observed by 8 hr. This ability is sustained in the mutant recipient for over 2 days when growth and division are inhibited. This restoration is not the trivial result of ion transfer. The membrane of the recipient mutant is profoundly changed, i.e., the inexcitable membrane is made excitable. This change has been documented by the return of Ca^{2+} action potential and action current to the mutant after injection (Fig. 3; Haga *et al.*, 1982).

Not only the cytoplasm from a wild-type cell is capable of this "curing." That of a mutant in a different complementation group from the recipient is just as potent. On the other hand, injection of cytoplasm from sister cells of the same mutant clone, or from different allelic variants in the same complementation group as that of the recipient mutant, always completely fails in the restoration. In other words, a mutant defective in one function but normal in a second function can help another mutant defective in the second function through microtransfusion just as in the case of classical genetic complementation. Results using the microtransfusion to classify pawns into different complementation groups are in complete agreement with the results from genetic crosses (Haga *et al.*, 1982). Furthermore, unlike fertilization, the injection pipette has no respect for species boundaries. It is therefore possible to take the cytoplasm from one species of *Paramecium* and to see whether it "cures" the mutant of a different species by injection. These interspecific complementation tests were used to see whether certain *cnr* genes are in fact the *pw* counterparts in a different species. All 18 combinations of cross-species transfusion using the three *pawn* and the three

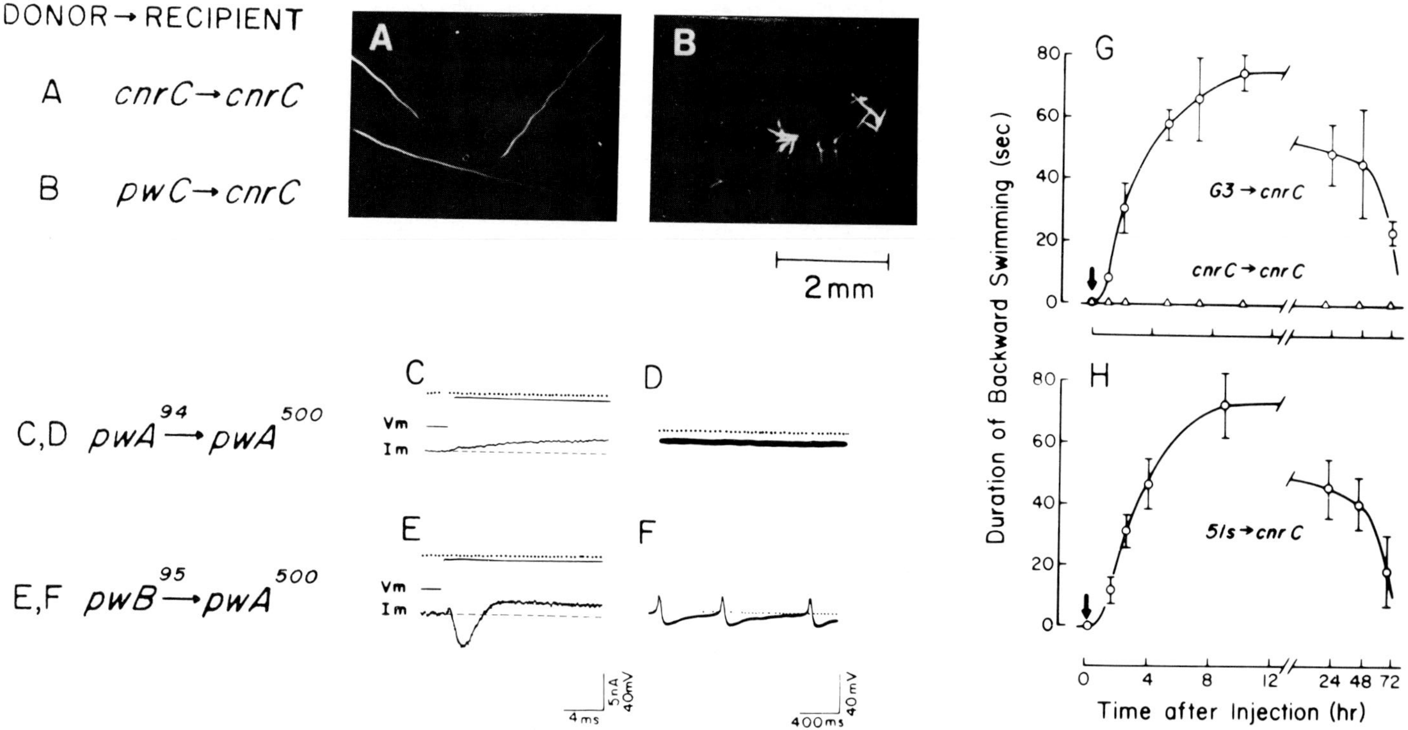
DONOR → RECIPIENT
A cnrC → cnrC
B pwC → cnrC
C,D pwA94 → pwA500
E,F pwB95 → pwA500
A
B
2mm
C
D
E
F
Vm
Im
5nA
40mV
4ms
40mV
400ms
G
H
G3 → cnrC
cnrC → cnrC
5/s → cnr C
Duration of Backward Swimming (sec)
Time after Injection (hr)
0
4
8
12
24
48
72
20
40
60
80

CNR mutants as donors or recipients effected curing, though of different degrees, i.e., any *pawn* complements any *CNR* and vice versa (Haga *et al.*, 1983). The simplest interpretation is that *pwA, pwB, pwC, cnrA, cnrB,* and *cnrC* are six different complementation groups existing in the genome of *P. tetraurelia* as well as *P. caudatum.* The positive complementations also mean that the structures and functions of the six gene products are conserved, at least over the evolutionary divergence of these two species.

Progress has been made toward purifying the factors in the wild-type cytoplasm, which cure the various mutants. Such purification starts with some 10 g wet weight with 1 g total protein, from 10^8 wild-type paramecia, axenically grown. Upon fractionation, the factors that restore excitability in the pawns (*pwA, pwB,* and *pwC*) are found to be in the microsomal fraction and are likely to be membrane proteins. The factor that cures the *cnrC* mutant, however, was found to be in a soluble fraction (Haga and Hiwatashi, 1982), which makes purification easier. Haga *et al.* (1984) have developed a procedure that enriches this factor over 3000-fold. This procedure includes homogenization of cells in the presence of protease inhibitors, differential centrifugations, ammonium sulfate fractionation, Sephadex G-75 gel filtration, hydroxylapatite column chromatography, and, finally, DEAE ion-exchange chromatography. As in an enzyme purification, each fraction has been tested for its curing activity by microinjection in the process of establishing this purification scheme. The activity, in sec/mg protein, is defined as the duration of backward swimming of *cnrC* recipients after being injected with the protein fraction. The backward swimming is induced by putting the cell in a K^+-rich depolarizing solution of

---

FIG. 3. "Curing" of inexcitable mutants by the injection of cytoplasm of wild-type or mutants of different complementation groups. (A) Behavior of the inexcitable mutant *cnrC* of *P. caudatum* 6–8 hr after receiving ~50 pl of cytoplasm from sister *cnrC* cells. Several injected recipients were collected from a rest solution and transferred to a solution of 4 m*M* Ba^{2+}, 1 m*M* Ca^{2+}, and 1 m*M* Tris, pH 7.2, and their reactions to this barium solution were registered by dark-field photography with a 3-sec exposure. The gentle helices indicate the usual forward swimming of the mutants that remained incapable of responding to the barium solution. (B) *cnrC* Injected with the cytoplasm of *pawnC,* a mutant of a different complementation group of a different species, *P. tetraurelia,* registered the clustered barbs due to the repeated jerks which are the consequences of the Ba^{2+}–Ca^{2+} action potentials. (C) The lack of Ca^{2+} action current and (D) the lack of Ba^{2+}–Ca^{2+} action potentials in a *pawnA* mutant injected with cytoplasm of another mutant of the same complementation group (*pawnA*). (E) The presence of the Ca^{2+} action current and (F) the presence of Ba^{2+}–Ca^{2+} action potentials in a *pawnA* mutant injected with cytoplasm of a mutant of a different complementation group (*pawnB*) (see Haga *et al.*, 1982, for details). (G,H) Time course of the restoration of excitability after microinjection of the cytoplasm. The recipients were injected at time 0 (arrows), periodically withdrawn from the rest solution, and temporarily transferred into Dryl's solution with 20 m*M* K^+. The full duration of backward swimming induced by this transfer, which is an estimate of membrane excitability, was measured and plotted over time. *cnrC* Mutants of *P. caudatum,* injected with the cytoplasm of G3, a wild-type *P. caudatum* (G), or injected with the cytoplasm of 51s, a wild-type *P. tetraurelia* (H), all show restorations, which peak by 8 hr after injection and last over 2 days (mean ± SD, n = 6) (see Haga *et al.*, 1983, for details).

fixed composition. This duration of backward swimming (in seconds) has been calibrated and found to be proportional to the maximal Ca^{2+} current induced by a depolarization under voltage clamp (Y. Satow, T. Hennessey, and N. Haga, unpublished results). The highly enriched fraction still contains many protein species. When the curing activity is correlated with protein band density, several candidates remain. Although the factor that restores excitability in the *cnrC* mutant has not been identified and purified to homogeneity, several of its properties are now known. It is clearly a protein since it is heat-labile, protease- and *N*-ethylmaleimide-sensitive, but RNase-resistant. It is soluble, acidic (p*I* 4.5–5.0), and small (less than 30,000 MW). Although these properties point in the direction of calmodulin, highly active curing fractions fail to stimulate brain phosphodiesterase activity (Haga *et al.*, 1984). It is of great interest that the function of a Ca^{2+} channel in the membrane is somehow dependent on a soluble factor in the cytoplasm. It brings to mind the observations that internal perfusion with buffer shuts off Ca^{2+} current in eggs and neurons (Kostyuk, 1980; Byerly and Hagiwara, 1982), and Ca^{2+} channels in a patch do not survive after the patch is detached from the rest of the neuron (Fenwick *et al.*, 1982; Hagiwara and Ohmori, 1983).

VI. *DANCER* MUTANTS, POSSIBLY DEFECTIVE IN Ca^{2+}-CHANNEL STRUCTURE

As stated above, we do not yet know whether any of the *pw* or the *cnr* genes code for the Ca^{2+} channel structure. Mutants of a new type called *dancer*, however, are good candidates for structural mutants. Their mutations, *Dn*, clearly change the characteristics of the Ca^{2+} current instead of erasing it (Hinrichsen and Saimi, 1984).

In the resting solution, the wild-type *Paramecium* generates Ca^{2+} action potentials graded to the stimuli, e.g., injected outward current. The *dancers* are more excitable; they give all-or-none action potentials upon stimulation (Fig. 4A and B). This explains their exaggerated avoiding reactions in certain solutions. Under voltage clamp, step depolarizations trigger inward Ca^{2+} currents in *dancers* that inactivate more slowly than those in the wild-type (Fig. 4C and D). The time constant of this Ca^{2+}-dependent Ca^{2+} channel inactivation during a step from -40 to -10 mV is about 1 msec for the wild-type but 3 msec for *dancer*. The inactivation in *dancers* is also far less complete. While the sustained current after the above inactivation is less than 0.5 nA in the wild-type, it is about 2 nA in *dancers*. A separate defect can be seen in the kinetics of deactivation of Ba^{2+} current, i.e., the tail inward current upon step repolarization is longer and slower in *dancer* than in the wild type. Passive properties of the membrane as well as other voltage-dependent currents are not affected by the *Dn* mutations. The poor inactivation of the Ca^{2+} channel in *dancers* results in a higher internal Ca^{2+} concentration. As expected, the Ca^{2+}-dependent K^+ current and the Ca^{2+}-

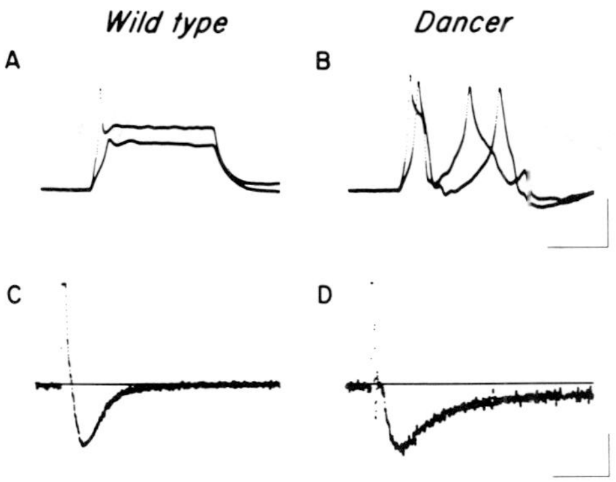

FIG. 4. Electrophysiological phenotypes of the *dancer* mutant. (A) Graded action potentials of the wild type induced by a 250-msec injected outward current of 0.4 or 1 nA. (B) The all-or-none action potentials of *dancer* by the identical current injections. Cells bathed in 4 m*M* K^+, 1 m*M* Ca^{2+}, 1 m*M* HEPES, 0.01 m*M* EDTA, pH 7.2–7.4. Recorded with 0.5 *M* KCl electrodes. (C) The transient Ca^{2+} current of the wild type induced by a step depolarization from −40 to −10 mV under a voltage clamp. The current rapidly inactivates within 10 msec after its peak. (D) The Ca^{2+} current of *dancer* inactivates slowly and incompletely. Cells bathed in 4 m*M* Cs^+, 1 m*M* Ca^{2+}, 10 m *M* TEA^+, 1 m*M* HEPES, and 0.01 m*M* EDTA, pH 7.2–7.4. Recorded with 4 *M* CsCl electrodes; calibration: (A,B) 125 msec, 20mV; (C,D) 5 msec, 5 nA (see Hinrichsen and Saimi, 1984, for details).

dependent Na^+ current are enhanced in the mutant. The simplest interpretation of the mutational alteration in the inactivation and deactivation processes is that the *Dn* mutation affects part of the channel structure causing functional abnormalities. The situation is reminiscent of the *shaker* mutational effects on the K^+ channel in *Drosophila* (Salkoff and Wyman, 1983) (see Jan *et al.*, Chapter 4, this volume).

All seven independently isolated *dancer* mutants are found to belong to the same complementation group. *Dn* is unlinked to *pwA, pwB, pwC,* and several other genes affecting different ion currents. Like the *shaker* mutation and other mutations thought to affect channel structures in *Drosophila, Dn* is codominant with its wild-type allele, Dn^+. Furthermore, quantitative analysis of the K^+-induced backward swimming after the proper genotypic changes shows that the establishment of the *dancer* phenotype and the establishment of the wild phenotype have a similar time course. This is very different from the situation with *pawns* where the wild phenotype is established almost immediately, whereas the full expression of the *pawn* phenotype takes several days after genotypic change. These results indicate that only a small amount of the pw^+ product is needed to support the full Ca^{2+} channel function, whereas the ratio of the Dn^+ product to the *Dn* product is directly proportional to the degree of normal function. The results are consistent with *pw* functioning in certain (enzymatic?) modifications and *Dn* coding for the channel itself. While *dancers* are mutants of *P. tetraurelia,* they bear some behavioral resemblance to the $cnrB^{ks}$ mutant of *P. caudatum.*

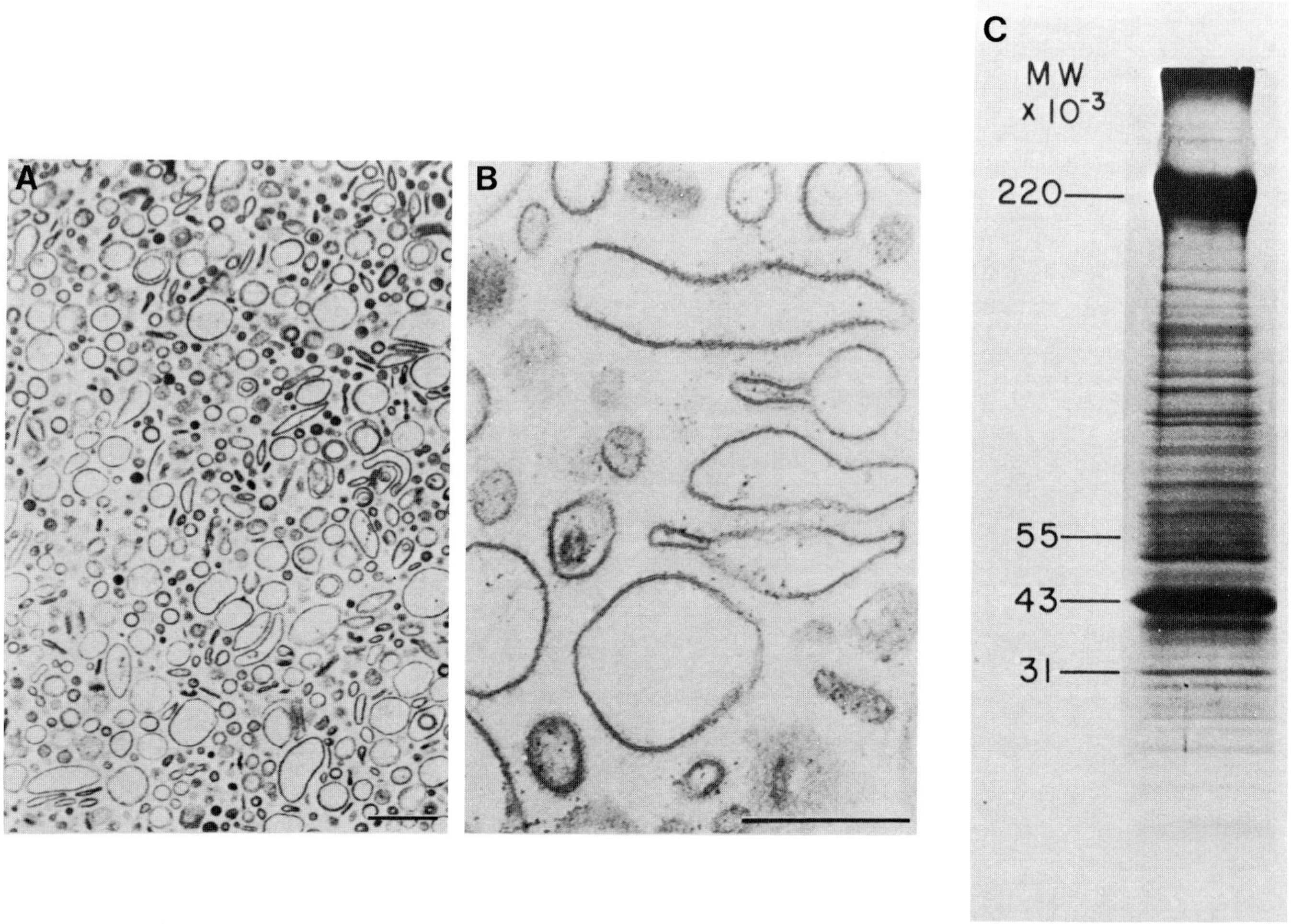
A
B
C
MW
x 10-3
220
55
43
31

VII. PURIFIED CILIARY MEMBRANE

To advance the study to the structure and function of ion channels *in vitro,* one needs to purify the membrane in which the channels are located. The Ca^{2+} channels, if not other channels, are located in the ciliary membrane of *Paramecium.* This membrane has been purified as follows. Paramecia (10^7) are collected and then deciliated with a "calcium shock" procedure. The bodies are discarded and the cilia collected through centrifugation. The cilia are vortexed in a buffer containing EDTA to remove the membrane from the axonemes. The material is then centrifuged on a sucrose gradient to separate the membrane at a sucrose step boundary. Judging by electron microscopic images and by gel bands, the membranes collected in this manner are free of axonemal or other contamination (Adoutte *et al.,* 1980).

The purified membranes are in vesicular form. Most of the vesicles are less than 1 μm in diameter and are unilamellar (Fig. 5A). The ciliary membrane is complex in protein composition. SDS–polyacrylamide gel electrophoreses separate more than 75 bands recognizable with Coomassie blue stain (Fig. 5C). The most abundant protein is a giant peptide of over 250,000 MW called the immobilization antigen, which accounts for over 70% of the total vesicle protein weight. Antisera against this protein immobilize the paramecium (Preer, 1959). *Paramecium* is capable of synthesizing many structurally and serologically distinct types of immobilization antigens, one type at a time in exclusion of the other types. The immobilization antigen forms a 100-Å-thick fuzzy layer covering the entire outer surface of the *Paramecium,* ciliary as well as somal. Mild protease treatment can strip off this layer (Ramanathan *et al.,* 1981). These major proteins, like the lipids, are probably part of the mechanical, electrical, and/or chemical environment in which the channels work. Polyclonal antibodies against the immobilization antigen can reduce or block completely the Ca^{2+} current (Eisenbach *et al.,* 1983; Ramanathan *et al.,* 1983), even when the antibodies are made monovalent and no longer immobilize the paramecia.

Although the protein species constituting the Ca^{2+} channels or other channels have not been identified, the membrane vesicles detached from *Paramecium* cilia apparently still contain some functional channels. Thiele and co-workers (1981, 1983) have enclosed arsenazo III inside the ciliary membrane vesicles and have shown that this Ca^{2+}-sensitive dye changes color when Ca^{2+} is added to the outside. At least part of this Ca^{2+} entry appears to be through functional Ca^{2+} channel since the vesicles prepared from several *pawn* mutants capture much less Ca^{2+}.

FIG. 5. Purified ciliary membranes and their proteins. Electron micrographs of purified ciliary membrane vesicles at lower (A) or higher (B) magnifications. Calibration bars are 0.25 μm. (C) SDS–polyacrylamide gel electrophoretogram of ciliary membrane proteins. The dominant high-molecular-weight protein is the immobilization antigen. The second most abundant protein(s) is about 42,000–44,000 MW (see Adoutte *et al.,* 1980, for details).

A different functional test for channel activities *in vitro* is to record the electric currents through them after they are put into an artificial lipid bilayer (Mueller and Rudin, 1969). When purified ciliary membrane of *Paramecium* is added to one of the two sides of a chamber separated by the bilayer and held at different voltages, step currents can be observed. At least two kinds of conductances are recognized: one with 30 pS, one with 2 pS. The larger conductance discriminates divalent from monovalent cations and the smaller conductance discriminates Ba^{2+} from Mg^{2+} and monovalents (Ehrlich *et al.*, 1983, 1984). Hanke *et al.* (1981) and Boheim *et al.* (1982) also reported single-channel currents after reassembly of ciliary membrane material into artificial lipid bilayer membrane. They even noticed a difference between the currents with wild-type material and those with *pawn* material, although it is not clear whether the currents are through the Ca^{2+} channel. Though preliminary, these results are very encouraging. Comparison between the macroscopic currents recorded *in vivo* with these microscopic currents registered *in vitro* will be most informative.

VIII. CONCLUSION

The use of a single-celled organism to study neurobiology may initially sound farfetched, but there are good reasons to believe it is not. As stated at the beginning of this article, we are interested in the very basic biochemical and biophysical characteristics used in sensory cells, nerves, and muscles, and we believe that these evolved early and are conserved in protozoa as well as in metazoa. The intricate axonemal structures, the intramembranous-particle arrays in various ciliary membranes, the Mg^{2+}-ATPase, the Ca^{2+} channel with its characteristic ion selectivity and inactivation mechanism, the voltage-dependent K^+ channels, the channels activated by internal Ca^{2+}, the action potential, and the use of Ca^{2+} as messenger are all identical or very similar in *Paramecium* and in metazoa. Thus, the universality of basic principles underlying behavior is far more than speculation. Some of these principles concern the structure, function, and regulation of ion channels in the membrane.

This article shows that it is relatively easy to isolate ion channel mutants of *Paramecium*, and a large collection of them is already in existence in two different species. These mutants will continue to be useful in sorting out various ion currents, microscopic or macroscopic, measured *in vivo* or *in vitro*. The *pw* and *cnr* mutants will continue to be used to subtract the Ca^{2+} current and its consequences, and *Dn* can be used to biologically inject more than the usual amount of Ca^{2+}. The mutations can also help us to identify the gene products important in ion channel function. The *cnrC* curing factor is a case in point.

A direct attack on the problem of channel structure and function would be to identify the genes coding for the channel proteins, clone them, sequence them, and perform site-specific mutageneses on them. Like the *shaker* of *Drosophila* (see Jan *et al.*, Chapter 4, this volume), *dancer* of *Paramecium* is a good candidate for being a structural mutant of a channel. If the Dn^+ product can eventually be identified and antibodies to it generated, it should be possible eventually to clone the *Dn* gene, if need be. Preer and co-workers (1981; Forney *et al.*, 1983), for example, have succeeded in cloning several sequences corresponding to the major membrane protein, the immobilization antigen of *Paramecium*. However, it is important to develop an *in vitro* assay for the ion channels. Otherwise, it would be very cumbersome to assess the functional activity of the products of the cloned putative structural genes for the channels. In this connection, the *in vitro* assay of the *Paramecium* channels reassembled into artificial lipid bilayers is a welcome development.

The ability to reassemble channels into an artificial membrane and show transmembrane current *in vitro* does not necessarily mean that the cytoplasm or other membrane proteins play no part in regulating the channel structure and function *in vivo*. Long-term modification of ion channels by phosphorylation, for example, is of great current interest. In the *Paramecium* system reviewed here, one wonders why a small acidic soluble protein (*cnrC* curing factor) is required for the function of the Ca^{2+} channel.

ACKNOWLEDGMENTS

We thank Drs. N. Bonini, A. Burgess-Cassler, T. Evans, M. Gustin, T. Hennessey, R. Hinrichsen, J. Kung, B. Martinac, D. L. Nelson, R. Ramanathan, E. Richard, and R.-H. Wu for critical readings of this article. Research for this article was supported by NSF BNS-8216149 and NIH PHS GM 22714.

REFERENCES

Adoutte, A., Ramanathan, R., Lewis, R. M., Dute, R. R., Ling, K.-Y., Kung, C., and Nelson, D. L. (1980). Biochemical studies of the excitable membrane of *Paramecium tetraurelia*. III. Proteins of cilia and ciliary membranes. *J. Cell Biol.* **84,** 717–729.

Adoutte, A., Ling, K.-Y., Forte, M., Ramanathan, R., Nelson, D. L., and Kung, C. (1981). Ionic channels of *Paramecium* from genetics and electrophysiology to biochemistry. *J. Physiol. (Paris)* **77,** 1145–1159.

Adoutte, A., Ling, K.-Y., Chang, S., Huang, F., and Kung, C. (1983). Physiological and mutational protein variations in the ciliary membrane of *Paramecium*. *Exp. Cell Res.* **148,** 387–404.

Berger, J. D. (1976). Gene expression and phenotypic change in *Paramecium tetraurelia* exconjugants. *Genet. Res.* **27,** 123–134.

Boheim, G., Hanke, W., Methfessel, C., Eibl, H., Kaupp, U. B., Maeliche, A., and Schultz, J. E. (1982). Membrane reconstitution below lipid phase transition temperature. *In* "Transport in Biomembranes: Model Systems and Reconstitution" (R. Antolini *et al.*, eds.), pp. 87–97. Raven, New York.

Brehm, P., and Eckert, R. (1978a). Calcium entry leads to inactivation of calcium channel in *Paramecium*. *Science* **202,** 1203–1206.

Brehm, P., and Eckert, R. (1978b). An electrophysiological study of the regulation of ciliary beating frequency in *Paramecium*. *J. Physiol. (London)* **283,** 557–568.

Brehm, P., Eckert, R., and Tillotson, D. (1980). Calcium-mediated inactivation of calcium current in *Paramecium*. *J. Physiol. (London)* **306,** 193–203.

Browning, J. L., and Nelson, D. L. (1976). Biochemical studies of the excitable membrane of *Paramecium aurelia*. I. $^{45}Ca^{2+}$ fluxes across resting and excited membrane. *Biochim. Biophys. Acta* **448,** 338–351.

Browning, J. L., Nelson, D. L., and Hansma, H. G. (1976). Ca^{2+} influx across the excitable membrane of behavioral mutants of *Paramecium. Nature* (*London*) **259,** 491–494.

Byerly, L., and Hagiwara, S. (1982). Calcium currents in internally perfused nerve cell bodies of *Limnea stagnalis. J. Physiol.* (*London*) **322,** 503–528.

Byrne, B. L., and Byrne, B. C. (1978). Behavior and the excitable membrane of *Paramecium. CRC Crit. Rev. Microbiol.* **6,** 53–103.

Chang, S. Y., and Kung, C. (1973a). Temperature-sensitive pawns: Conditional behavioral mutants of *Paramecium aurelia. Science* **180,** 1197–1199.

Chang, S. Y., and Kung, C. (1973b). Genetic analyses of heat sensitive pawn mutants of *Paramecium aurelia. Genetics* **75,** 49–59.

Chang, S. Y., Van Houten, J., Robles, L., Lui, S., and Kung, C. (1974). An extensive behavioral and genetic analysis of the Pawn mutants of *Paramecium aurelia. Genet. Res.* **23,** 165–173.

Cronkite, D. L. (1979). The genetics of swimming behavior and mating behavior in *Paramecium. In* "Biochemistry and Physiology of Protozoa" (S. H. Hunter and M. Levandowsky, eds.), Vol. 2, pp. 222–275. Academic Press, New York.

DiNallo, M. C., Wohlford, M., and Van Houten, J. (1982). Mutants of *Paramecium* defective in chemokinesis to folate. *Genetics* **102,** 149–158.

Doughty, M. J., and Dryl, S. (1981). Control of ciliary activity in *Paramecium:* An analysis of chemosensory transduction in a unicellular eukaryotic organism. *Prog. Neurobiol.* **16,** 1–115.

Dunlap, K. (1977). Localization of calcium channels in *Paramecium caudatum. J. Physiol.* (*London*) **271,** 119–133.

Eckert, R., and Brehm, P. (1979). Ionic mechanisms of excitation in *Paramecium. Annu. Rev. Biophys. Bioeng.* **8,** 353–383.

Eckert, R., Naitoh, Y., and Friedman, K. (1972). Sensory mechanism in *Paramecium.* I. Two components of the electric response to mechanical stimulation of the anterior surface. *J. Exp. Biol.* **546,** 683–694.

Eckert, R., Naitoh, Y., and Machemer, H. (1976). Calcium in the bioelectric and motor functions of *Paramecium. In* "Calcium in Biological Systems" (C. J. Duncan, ed.), pp. 233–255. Cambridge Univ. Press, London and New York.

Ehrlich, B. E., Finkelstein, A., Forte, M., and Kung, C. (1983). Calcium channels from *Paramecium* cilia incorporated in a planar lipid bilayer. *Biophys. J.* **41,** 293A.

Ehrlich, B. E., Finkelstein, A., Forte, M., and Kung, C. (1984). Voltage-dependent calcium channels from *Paramecium* cilia incorporated into planar lipid bilayers. *Science* **225,** 427–428.

Eisenbach, L., Ramanathan, R., and Nelson, D. L. (1983). Biochemical studies of the excitable membrane of *Paramecium tetraurelia.* IX. Antibodies against ciliary membrane proteins. *J. Cell Biol.* **97,** 1412–1420.

Fenwick, E. M., Marty, A., and Neher, E. (1982). Sodium and calcium channels in bovine chromaffin cells. *J. Physiol.* (*London*) **331,** 599–635.

Forney, J. D., Epstein, L. M., Preer, L. B., Rudman, B. M., Widmayer, D. J., Klein, W. H., and Preer, J. R., Jr. (1983). Structure and expression of genes for surface proteins in *Paramecium. Mol. Cell Biol.* **3,** 466–474.

Forte, M., Satow, Y., Nelson, D., and Kung, C. (1981). Mutational alteration of membrane phospholipid composition and voltage-sensitive ion channel function in *Paramecium. Proc. Natl. Acad. Sci. U.S.A.* **78,** 7195–7199.

Gibbons, I. R. (1981). Cilia and flagella of eukaryotes. *J. Cell Biol.* **91,** 107s–124s.

Gustin, M. C., Bonini, N. M., and Nelson, D. L. (1983). Membrane potential regulation of cAMP: control mechanism of swimming behavior in the ciliate *Paramecium. Neurosci. Abstr.* **9,** 167.

Haga, N., and Hiwatashi, K. (1982). A soluble gene product controlling membrane excitability in *Paramecium*. *Cell Biol. Int. Rep.* **6,** 295–300.

Haga, N., Forte, M., Saimi, Y., and Kung, C. (1982). Microinjection of cytoplasm as a text of complementation in *Paramecium*. *J. Cell Biol.* **82,** 559–564.

Haga, N., Saimi, Y., Takahashi, M., and Kung, C. (1983). Intra- and interspecific complementation of membrane-inexcitable mutants of *Paramecium*. *J. Cell Biol.* **97,** 378–382.

Haga, N., Forte, M., Ramanathan, R., Saimi, Y., Takahashi, M., and Kung, C. (1984). Purification of a soluble protein controlling Ca^{2+} channel activity in *Paramecium*. *Biophys. J.* **45,** 130–132.

Hagiwara, S., and Ohmori, H. (1983). Studies of a single calcium channel current in rat clonal pituitary cells. *J. Physiol. (London)* **336,** 649–661.

Hanke, W., Eibl, H., and Boheim, G. (1981). A new method for membrane reconstitution: Fusion of protein-containing vesicles with planar bilayer membranes below lipid phase transition temperature. *Biophys. Struct. Mech.* **7,** 131–137.

Hansma, H., and Kung, C. (1976). Defective ion regulation in a class of membrane-excitation mutants in *Paramecium*. *Biochem. Biophys. Acta* **463,** 128–139.

Hennessey, T. M., and Kung, C. (1984). An anticalmodulin drug, W-7, inhibits the voltage-dependent calcium current in *Paramecium caudatum*. *J. Exp. Biol.* **110,** 169–181.

Hennessey, T., and Nelson, D. L. (1979). Thermosensory behavior in *Paramecium tetraurelia:* A quantitative assay and some factors that influence thermal avoidance. *J. Gen. Microbiol.* **112,** 337–357.

Hennessey, T. M., Saimi, Y., and Kung, C. (1983). A heat-induced depolarization of *Paramecium* and its relationship to thermal avoidance behavior. *J. Comp. Physiol.* **153,** 39–46.

Hinrichsen, R. D., and Kung, C. (1984). Genetic analysis of axonemal mutants in *Paramecium tetraurelia* defective in their response to calcium. *Genet. Res.* **43,** 11–20.

Hinrichsen, R. D., and Saimi, Y. (1984). A mutation that alters properties of the Ca^{2+} channel in *Paramecium tetraurelia*. *J. Physiol. (London)* **351,** 397–410.

Hinrichsen, R. D., Saimi, Y., Hennessey, T., and Kung, C. (1984a). Mutants in *Paramecium tetraurelia* defective in their axonemal response to calcium. *Cell Motility* **4,** 283–295.

Hinrichsen, R. D., Saimi, Y., and Kung, C. (1984b). Mutants with altered Ca^{++}-channel properties in *Paramecium tetraurelia*: Isolation, characterization and genetic analysis. *Genetics* (in press).

Hiwatashi, K. (1969). Paramecium. *In* "Fertilization II" (C. B. Metz and A. Monroy, eds.), pp. 255–293. Academic Press, New York.

Hiwatashi, K., Haga, N., and Takahashi, M. (1980). Restoration of membrane excitability in a behavioral mutant of *Paramecium caudatum* during conjugation and by microinjection of wild-type cytoplasm. *J. Cell Biol.* **84,** 476–480.

Jennings, H. S. (1906). "Behavior of Lower Organisms." Indiana Univ. Press, Bloomington, Indiana.

Kinosita, H., Dryl, S., and Naitoh, Y. (1964). Changes in membrane potential and the responses to stimuli in *Paramecium*. *J. Fac. Sci. Univ. Tokyo, Sect. IV* **10,** 291–301.

Kostyuk, P. G. (1980). Calcium ionic channels in electrically excitable membrane. *Neuroscience* **5,** 945–959.

Kung, C. (1971a). Genic mutations with altered system of excitation in *Paramecium aurelia*. II. Mutagenesis, screening and genetic analysis of the mutants. *Genetics* **69,** 29–45.

Kung, C. (1971b). Genic mutations with altered system of excitation in *Paramecium aurelia*. I. Phenotypes of the behavioural mutants. *Z. Vergl. Physiol.* **71,** 142–164.

Kung, C. (1976). Membrane control of ciliary motions and its genetic modification. *In* "Cell Motility" (R. Goldman, T. Pollard, and J. Rosenbaum, eds.) Book C, pp. 941–948. Cold Spring Harbor Laboratory, Cold Spring Harbor, New York.

Kung, C. (1979). The biology and genetics of paramecium behavior. *In* "Topics in Neurogenetics" (X. O. Breakefield, ed.), pp. 1–26. Elsevier, Amsterdam.

Kung, C., and Eckert, R. (1972). Genetic modifications of electric properties in an excitable membrane. *Proc. Natl. Acad. Sci. U.S.A.* **69,** 93–97.

Kung, C., and Naitoh, Y. (1973). Calcium-induced ciliary reversal in the extracted models of "Pawns", a behavioral mutant of *Paramecium. Science* **179,** 195–196.

Kung, C., and Saimi, Y. (1982). The physiological basis of taxes in *Paramecium. Annu. Rev. Physiol.* **44,** 519–534.

Ling, K.-Y., and Kung, C. (1980). Ba^{2+} influx measures the duration of membrane excitation in *Paramecium. J. Exp. Biol.* **84,** 73–87.

Machemer, H., and de Peyer, J. (1977). Swimming sensory cells. Electrical membrane parameters, receptor properties and motor control in ciliated protozoa. *Verh. Dtsch. Zool. Ges.* pp. 86–110. Fischer, Stuttgart.

Machemer, H., and Ogura, A. (1979). Ionic conductances of membranes in ciliated and deciliated *Paramecium. J. Physiol. (London)* **296,** 49–60.

Merkel, S., Kaneshiro, E., and Gruenstein, E. I. (1981). Characterization of the cilia and ciliary membrane proteins of wild-type *Paramecium tetraurelia* and a pawn mutant. *J. Cell Biol.* **89,** 206–215.

Mueller, P., and Rudin, D. O. (1969). Bimolecular lipid membranes. Techniques of formation, study of electrical properties, and induction of gating phenomena. *In* "Laboratory Techniques in Membrane Biophysics" (H. Passow and R. Stämpfli, eds.), pp. 141–156. Springer-Verlag, Berlin and New York.

Naitoh, Y. (1974). Bioelectric basis of behavior in protozoa. *Am. Zool.* **14,** 883–893.

Naitoh, Y., and Eckert, R. (1968). Electrical properties of *Paramecium caudatum:* Modification by bound and free cations. *Z. Vergl. Physiol.* **61,** 427–452.

Naitoh, Y., and Eckert, R. (1969). Ionic mechanisms controlling behavioral responses of *Paramecium* to mechanical stimulation. *Science* **164,** 963–965.

Naitoh, Y., and Eckert, R. (1972). Electrophysiology of ciliate protozoa. *Exp. Physiol. Biochem.* **5,** 17–31.

Naitoh, Y., and Eckert, R. (1973). Sensory mechanisms in *Paramecium*. II. Ionic basis of the hyperpolarizing mechanoreceptor potential. *J. Exp. Biol.* **59,** 53–65.

Naitoh, Y., and Kaneko, H. (1972). Reactivated triton-extracted models of *Paramecium:* Modification of ciliary movement by calcium ions. *Science* **176,** 523–524.

Nelson, D. L., and Kung, C. (1978). Behavior of *Paramecium:* Chemical, physiological and genetic studies. *Recept. Recognition Ser. B* **5,** 75–100.

Oertel, D., Schein, S. J., and Kung, C. (1977). Separation of membrane currents using a *Paramecium* mutant. *Nature (London)* **268,** 120–124.

Oertel, D., Schein, S. J., and Kung, C. (1978). A potassium conductance activated by hyperpolarization in *Paramecium. J. Membr. Biol.* **43,** 169–185.

Ogura, A., and Machemer, H. (1980). Distribution of mechanorecepor channels in the *Paramecium* surface membrane. *J. Comp. Physiol.* **135,** 233–242.

Ogura, A., and Takahashi, K. (1976). Artificial deciliation causes loss of calcium-dependent responses in *Paramecium. Nature (London)* **264,** 170–172.

Omoto, C. K., and Kung, C. (1979). The pair of central tubules rotates during ciliary beat in *Paramecium. Nature (London)* **279,** 532–534.

Otter, Y., and Salmon, E. D. (1979). Hydrostatic pressure reversibly blocks membrane control of ciliary motility in *Paramecium. Science* **206,** 358–361.

Preer, J. R., Jr. (1959). Studies on the immobilization antigens of *Paramecium*. III. Properties. *J. Immunol.* **83,** 385–391.

Preer, J. R., Preer, L. B., and Rudman, B. M. (1981). mRNAs for the immobilization antigens of *Paramecium*. *Proc. Natl. Acad. Sci. U.S.A.* **78,** 6776–6778.

Ramanathan, R., Adoutte, A., and Dute, R. (1981). Biochemical studies of the excitable membrane of *Paramecium tetraurelia*. V. Effects of proteases on the ciliary membrane. *Biochim. Biophys. Acta* **641,** 349–365.

Ramanathan, R., Saimi, Y., Peterson, J. B., Nelson, D. L., and Kung, C. (1983). Antibodies to the ciliary membrane of *Paramecium tetraurelia* alter membrane excitability. *J. Cell Biol.* **97,** 1421–1428.

Saimi, Y., and Kung, C. (1980). A Ca-induced Na^+ current in *Paramecium*. *J. Exp. Biol.* **88,** 305–325.

Saimi, Y., and Kung C. (1982). Are ions involved in the gating of calcium channels? *Science* **218,** 153–156.

Saimi, Y., Hinrichsen, R. D., Forte, M., and Kung, C. (1983). Mutant analysis shows that the Ca^{2+}-induced K^+ current shuts off one type of excitation in *Paramecium*. *Proc. Natl. Acad. Sci. U.S.A.* **80,** 5112–5116.

Salkoff, L., and Wyman, R. (1983). Ion channels in *Drosophila* muscle. *Trends Neurosci.* **5,** 128–133.

Satow, Y. (1978). Internal calcium concentration and potassium permeability in *Paramecium*. *J. Neurobiol.* **9,** 81–91.

Satow, Y., and Kung, C. (1974). Genetic dissection of active electrogenesis in *Paramecium aurelia*. *Nature (London)* **247,** 69–71.

Satow, Y., and Kung, C. (1976a). Mutants with reduced Ca activation in *Paramecium aurelia*. *J. Membr. Biol.* **28,** 277–294.

Satow, Y., and Kung, C. (1976b). A TEA^+-insensitive mutant with increased potassium conductance in *Paramecium aurelia*. *J. Exp. Biol.* **65,** 51–63.

Satow, Y., and Kung, C. (1976c). A mutant of *Paramecium* with increased relative resting potassium permeability. *J. Neurobiol.* **7,** 325–338.

Satow, Y., and Kung, C. (1979). Voltage-sensitive Ca-channels and the transient inward current in *Paramecium tetraurelia*. *J. Exp. Biol.* **78,** 149–161.

Satow, Y., and Kung, C. (1980a). Ca-induced K^+-outward current in *Paramecium tetraurelia*. *J. Exp. Biol.* **88,** 293–303.

Satow, Y., and Kung, C. (1980b). Membrane currents of pawn mutants of the *pwA* group in *Paramecium tetraurelia*. *J. Exp. Biol.* **84,** 57–71.

Satow, Y., and Kung, C. (1981). Possible reduction of surface charge by a mutation. *J. Membr. Biol.* **59,** 179–190.

Satow, Y., Chang, S. Y., and Kung, C. (1974). Membrane excitability: Made temperature dependent by mutations. *Proc. Natl. Acad. Sci. U.S.A.* **71,** 2703–2706.

Satow, Y., Hansma, H. G., and Kung, C. (1976). The effect of sodium on "paranoiac"—a membrane mutant of *Paramecium*. *Comp. Biochem. Physiol.* **54A,** 323–329.

Satow, Y., Murphy, A. D., and Kung, C. (1983). The ionic basis of the depolarizing mechanoreceptor potential of *Paramecium tetraurelia*. *J. Exp. Biol.* **103,** 253–264.

Schein, S. J. (1976a). Nonbehavioral section for pawns, mutants of *Paramecium aurelia* with decreased excitability. *Genetics* **84,** 452–468.

Schein, S. J. (1976b). Calcium channel stability measured by gradual loss of excitability in Pawn mutants of *Paramecium aurelia*. *J. Exp. Biol.* **65,** 725–736.

Schein, S. J., Bennett, M. V. L., and Katz, G. M. (1976). Altered calcium conductance in pawns, behavioral mutants of *Paramecium aurelia*. *J. Exp. Biol.* **65,** 699–724.

Shusterman, C. L., Thiede, E. W., and Kung, C. (1978). K^+-resistant mutants and "adaptations" in *Paramecium*. *Proc. Natl. Acad. Sci. U.S.A.* **75,** 5645–5649.

Sonneborn, T. M. (1970). Methods in *Paramecium* research. *Methods Cell Physiol.* **4,** 243–339.

Sonneborn, T. M. (1975). *Paramecium aurelia. Handb. Genet.* **2,** 469–594.

Takahashi, M. (1979). Behavioral mutants in *Paramecium caudatum. Genetics* **92,** 393–408.

Takahashi, M., and Naitoh, Y. (1978). Behavioral mutants of *Paramecium caudatum* with defective membrane electrogenesis. *Nature* (*London*) **271,** 656–658.

Thiele, J., and Schultz, J. E. (1981). Ciliary membrae vesicles of *Paramecium* contain the voltage-sensitive calcium chanel. *Proc. Natl. Acad. Sci. U.S.A.* **78,** 3688–3691.

Thiele, J., Otto, M. K., Dietmer, J. W., and Schultz, J. E. (1983). Calcium channels of the excitable ciliary membrane from *Paramecium:* An initial biochemical characterization. *J. Membr. Biol.* **76,** 253–260.

Van Houten, J. (1977). A mutant of *Paramecium* defective in chemotaxis. *Science* **198,** 745–748.

Van Houten, J. (1979). Membrane potential changes during chemokinesis in *Paramecium. Science* **204,** 1100–1103.

Van Houten, J. (1981). Chemosensory transduction in *Paramecium:* Role of membrane potential. *Olfact. Taste* **7,** 53–56.

Van Houten, J., Hansma, H. G., and Kung, C. (1975). Two quantitative assays for chemotaxis in *Paramecium. J. Comp. Physiol.* **104,** 211–223.

Van Houten, J., Chang, S. Y., and Kung, C. (1977). Genetic analysis of "Paranoiac" mutants of *Paramecium aurelia. Genetics* **86,** 113–120.

Van Houten, J., Hauser, D. C. R., and Levandowsky, M. (1981). Chemosensory behavior in Protozoa. *Biochem. Physiol. Protozoan* **4,** 67–124.

Chapter 4

Studies of *Shaker* Mutations Affecting a K$^+$ Channel in *Drosophila*

LILY YEH JAN, SANDRA BARBEL,* LESLIE TIMPE,* CHERYL LAFFER,* LAWRENCE SALKOFF,‡ PATRICK O'FARRELL,† AND YUH NUNG JAN**

Departments of Physiology and Biochemistry†*
University of California
San Francisco, California

and

‡Department of Anatomy and Neurobiology
Washington University School of Medicine
St. Louis, Missouri

I. INTRODUCTION

Several different types of voltage-sensitive potassium channels have been characterized in recent biophysical studies (Adams *et al.*, 1980). Different neurons have been shown to have rather different makeup of these K$^+$ channels, which to a large extent determines the firing pattern and shape of action poten-

ISBN 0-12-153323-9

tials in neurons (Byrne, 1980). Since the amount of transmitter released from a neuron can be drastically altered by slight alterations of the shape or duration of the action potential (Klein and Kandel, 1980; Llinas *et al.*, 1982), conceivably K^+ channels may play an important role in the control of synaptic efficacy. In fact, some K^+ channels are known to be influenced not only by membrane potential but also by transmitter molecules such as acetylcholine or serotonin and intracellular messengers such as Ca^{2+} or cAMP (e.g., Adams *et al.*, 1982; DePeyer *et al.*, 1982). For instance, the mechanism underlying sensitization in *Aplysia* appears to involve the inactivation of a particular K^+ channel by serotonin (Klein *et al.*, 1982; Siegelbaum *et al.*, 1982). To understand better how the expression of different K^+ channels is regulated in neurons and how the activity of individual K^+ channels may be controlled by various chemical messengers, one needs to characterize K^+ channels in molecular terms. Unfortunately, little information is available concerning the structure of K^+ channels or their genes because there are no antibodies against K^+ channels and, with the exception of Ca^{2+}-activated K^+ channels (Hugues *et al.*, 1982), no toxins with high affinity have been found for K^+ channels.

One possible approach to a molecular study of K^+ channels is to clone the genes for K^+ channels. This involves the traditional genetic approach for identifying genes that code for protein molecules. It can be done with animal systems, such as *Drosophila,* for which genetic tools are well developed. The first step in this approach is the induction of a mutation which gives a clear indication of affecting a K^+ channel. The mutant gene that is shown to alter normal K^+ conductance by physiological tests can be mapped genetically to a particular locus. Once the locus is known the gene can then be cloned by a variety of techniques. In *Drosophila,* cloning may be done in a brute force manner by "walking" along the chromosome. Alternatively, transposable elements may be used to "tag" the gene and facilitate cloning (Bingham *et al.*, 1981; Modolell *et al.*, 1983; Searles *et al.*, 1982). Transposable elements are DNA fragments that apparently can generate and insert copies of themselves elsewhere in the genome, often causing mutations at the sites of insertion. The frequency of transposition of a particular transposable element, the p factor, is drastically increased by mating appropriate strains of fruit flies (Engels, 1981; Rubin *et al.*, 1982) (see Section II). This "hybrid dysgenesis" phenomenon may then be used systematically for the cloning of genes in *Drosophila,* as outlined in Fig. 1.

In this article, we shall first review evidence that the *Shaker* locus in *Drosophila* is the site for the structural gene of a K^+ channel and discuss the strategies to be used for the molecular cloning of the *Shaker* locus. Then, we shall describe the recently isolated hybrid dysgenesis-induced *Shaker* mutants (Sh^{SBn}, $n = 1, 2 \ldots 13$; isolated by Sandra Barbel), which may be useful in the initial cloning and subsequent analysis of DNA from the *Shaker* locus.

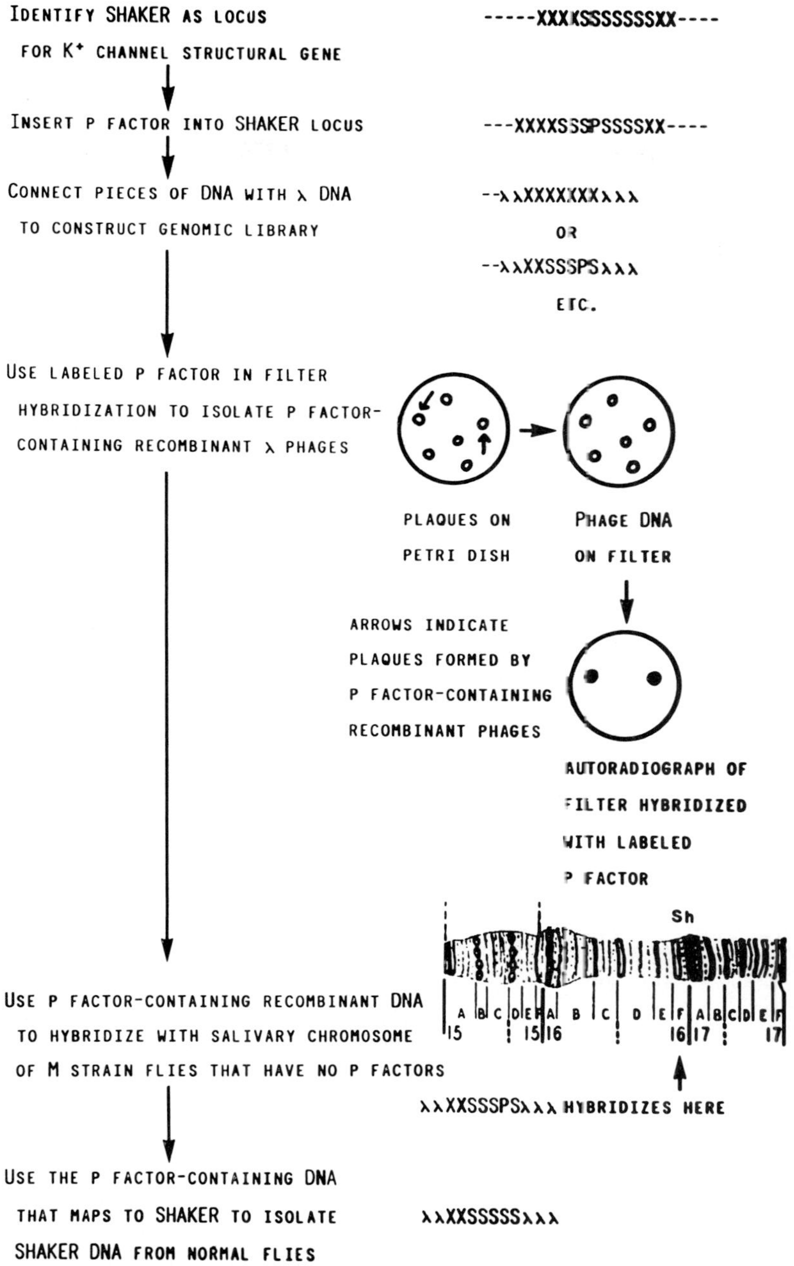

FIG. 1. Strategies for cloning DNA from the *Shaker* locus. As detailed in the text, identification of *Shaker* as a possible locus for K^+-channel structural gene was done using a combination of electrophysiological and genetic studies. Hybrid dysgenesis was used to induce the insertion of p factors into the *Shaker* locus. Techniques have been well worked out for construction of genomic libraries (Maniatis *et al.*, 1978), filter hybridization (Benton and Davis, 1977), and *in situ* hybridization with salivary chromosome (Pardue and Gall, 1975). P, Transposable p factor; S, sequences within the *Shaker* locus; X, flanking sequences on the X chromosome; and λ, DNA of the bacteriophage.

II. BACKGROUND

The *Shaker* Locus Probably Contains the Structural Gene for a K^+ Channel—A Review

1. Identification of *Shaker* as a Locus Important for K^+ Channel Function

Mutations affecting nervous function most likely would cuase behavioral abnormality, paralysis, or lethality. With this expectation, over 100 existing behavioral mutants were screened electrophysiologically for possible defects in larval neuromuscular transmission. From that search, mutations of the *Shaker* locus were found to cause abnormally large and prolonged transmitter release, especially when the extracellular Ca^{2+} concentration was low (Jan *et al.*, 1977). Apparently the nerve action potentials in *Shaker* mutants were prolonged and carried recurrent spikes (Jan and Jan, 1980). This caused prolonged elevation of calcium conductance at the nerve terminal and prolonged transmitter release (Jan *et al.*, 1977). The *Shaker* phenotype was attributed to a defect in K^+ channel function, because the behavioral and physiological phenotype of *Shaker* can be mimicked by treating wild-type larvae with 4-aminopyridine (4-AP), a K^+ channel blocker, whereas blocking Na^+ or Ca^{2+} channels did not correct for the abnormality (Jan *et al.*, 1977). Recent voltage clamp studies of *Drosophila* pupal muscles showed that 4-AP specifically blocked one type of K^+ channel, the A channel (Salkoff, 1983). Moreover, various *Shaker* mutations were found to affect the A channel function specifically (Salkoff, 1983), as discussed in subsection 2.

2. Evidence That the *Shaker* Locus Contains the Structural Gene for a K^+ Channel

In addition to abnormal synaptic transmission, *Shaker* mutants showed prolonged action potential duration in the cervical giant fibers of adult flies (Tanouye *et al.*, 1981), as well as altered A current in pupal muscles (Salkoff and Wyman, 1981). These studies have raised the possibility that the *Shaker* locus contains the structural gene for the A channel. This section summarizes the existing evidence.

a. The Null Phenotype Caused by Deletion in Shaker *Locus.* Deleting a small fragment within the *Shaker* locus [between the breakpoints of T(X;Y) B55 and T(X;Y) W32] eliminates the A current in pupal muscles (Salkoff, 1983) and results in prolonged transmitter release from larval nerve terminals, similar to that found in the Sh^{KS133} mutant. This suggests that genes in the *Shaker* locus are necessary for the expression of functional A channels.

b. Gene Dosage Studies. Although the Sh^{KS133} mutant exhibits the same phenotype as the deletion ($B55^{D}/W32^{P}$) (the null phenotype), the mutation apparently causes the production of an abnormal gene product because the Shaker phenotype is not totally suppressed even in flies carrying enough copies of the Sh^{+} gene to make close to 100% of the normal gene product (Tanouye *et al.*, 1981; Salkoff, 1983). This indicates that the gene affected by the Sh^{KS133} mutation is a structural gene.

c. Altered Voltage Dependence of A Channel Inactivation in the Sh^5 *Mutant.* Study of a less severe *Shaker* mutant, Sh^5, has provided additional evidence for this hypothesis (Salkoff, 1983). In pupal muscle of Sh^5 flies, the A current has normal amplitude and rise time, but it inactivates more abruptly than in wild-type. While normally the time constant for inactivation shows strong voltage dependence, in Sh^5 it is small and independent of membrane potential. Although the phenotype of Sh^5 differs from that of Sh^{KS133}, the two mutations are probably allelic; no recombination was observed between these two mutations among 8000 flies scored (Salkoff, 1983). The fact that one mutation of the *Shaker* locus alters the voltage dependence of A channel inactivation, whereas another mutation eliminates the A current, suggests that the *Shaker* locus contains the structural gene of the A channel.

3. Use of Transposable Elements in the Cloning of Genes in *Drosophila*

The cloning of genes whose product is either not defined or not available because of difficulties in isolation can represent a formidable problem. In *Drosophila* two general approaches have been developed to isolate the DNA sequences of genes that have been defined only by genetic criteria.

Most frequently these genes have been cloned by first genetically defining the position of the locus of interest on the polytene chromosome map and then obtaining a nucleic acid sequence (usually a piece of cloned DNA or an mRNA) that hybridizes to the polytene chromosomes at a cytological position that is near the locus of interest. Using a process referred to as "chromosome walking," it is possible to isolate adjacent cloned sequences overlapping the start clone and by a reiterative process "walk down the chromosome" to the region of interest. This process is often exceedingly laborious and because of the inaccuracies of cytological localization the length of a walk cannot be reliably predicted.

An alternative proposed by Bingham *et al.* (1981) has been called "transposon tagging." Should transposition lead to a damaging insertion within a gene, the mutant gene sequences will be "tagged" by the presence of the transposable element. If the transposable element DNA has been isolated, clones of tagged sequences can be identified by hybridization between the linked tag and labeled transposable element DNA. This approach has been used to clone the genes for *white, scute,* and *RNA polymerase* (Bingham *et al.*, 1981; Modolell *et al.*, 1983; Searles *et al.*, 1982).

The p element of *Drosophila* is a transposon that is particularly useful for transposon tagging because its frequency of transposition can be induced to levels that result in useful frequencies of transposon-induced mutations. Transposition is drastically increased by crosses between males from a strain containing p elements (P strain) with females from a strain lacking p elements (M strain) (Engels, 1981; Rubin *et al.*, 1982). The progeny of these "dysgenic crosses" can be screened for any particular mutant phenotype, and these will usually be due to a damaging insertion of p element sequences into the gene of interest. If a bank of clones is produced from the new mutant, those clones carrying tagged sequences can be identified by hybridization to labeled p element DNA (see Fig. 1).

III. HYBRID DYSGENESIS-INDUCED *SHAKER* MUTANTS

Knowing that the *Shaker* locus is likely to contain the structural gene for the A channel (see Section II), we undertook the isolation of DNA from the *Shaker* locus for molecular studies (Jan *et al.*, 1983). The strategy used was to first isolate *Shaker* mutations induced by hybrid dysgenesis, presumably caused by the insertion of p factors into the *Shaker* locus (see Section II). With such mutants we used labeled p factor to isolate DNA that is contiguous with the p factor within the *Shaker* locus.

A. The Physiological Abnormalities of Dysgenesis-Induced Sh^{SBn} Are Similar to Those of Existing *Shaker* Mutants

Having found several dysgenesis-induced mutants that shake under ether anesthesia, we first wanted to know whether they showed physiological abnormalities characteristic of *Shaker* mutants, namely, an altered A current and/or prolonged transmitter release caused by prolonged nerve action potential. We found that some Sh^{SBn} mutants showed the extreme, null phenotype of *Shaker,* whereas other dysgenesis-induced Sh^{SBn} mutants showed a weaker *Shaker* phenotype (Jan *et al.*, 1983).

B. Sh^{SBn} Mutations Appear to Affect the *Shaker* Locus

Knowing that the Sh^{SBn} mutations caused physiological abnormalities similar to those of *Shaker* mutants, we then asked whether these various Sh^{SBn} mutations mapped to the same region on the X chromosome as the *Shaker* locus.

Using both genetic recombination and complementation tests involving chromosomal rearrangements which result in the duplication or deletion of the fragment of X chromosome containing the *Shaker* locus, we mapped all dysgenesis-induced Sh^{SBn} mutations roughly to the *Shaker* locus (Jan *et al.*, 1983).

C. A p Factor Is Present in or near the *Shaker* Locus of Most Sh^{SBn} Mutants

The purpose of generating *Shaker* mutations via hybrid dysgenesis was to insert a p factor into the *Shaker* locus so that the *Shaker* DNA sequences tagged with the p factor could be identified and isolated by hybridization with the labeled p element (see Section II). Having isolated and mapped dysgenesis-induced mutations that showed the typical *Shaker* phenotype, we then did *in situ* hybridization experiments to test whether the radioactive p element probe would hybridize to the 16F region of the salivary chromosome, which contained the *Shaker* locus (Tanouye *et al.*, 1981).

The radioactive p element probe was made from a recombinant plasmid, p 25.1, provided by G. M. Rubin. Radioactive label was found around the 16F region in several Sh^{SBn} mutants examined, indicating that a p factor is inserted into or near the *Shaker* locus in these *Shaker* mutants (Jan *et al.*, 1983).

To summarize, we have isolated *Shaker* mutants through dysgenic crosses which promoted the transposition of p elements. These mutations showed both behavioral and physiological phenotypes of *Shaker* mutations and mapped to roughly the region of the *Shaker* locus. Finally, a p factor was found in or near the *Shaker* locus by *in situ* hybridization with the salivary chromosomes. These results suggest that the Sh^{SBn} mutations were induced by the insertion of a p factor. If we have indeed tagged the *Shaker* locus with a p element, these Sh^{SBn} mutations may prove very useful in the cloning and molecular analysis of the *Shaker* locus.

IV. DISCUSSION

Molecular studies of K^+ channels are important because these channels are likely to play important roles in the control of neuronal activity and synaptic efficacy. The genetics of *Drosophila* and mutations of the *Shaker* locus offer an alternative approach for cloning K^+ channels in the absence of high-affinity toxins or antibodies against K^+ channels. Genetic analyses using electrophysiological assays including voltage clamping have provided strong evidence that the *Shaker* locus contains the structural gene for a K^+ channel, the A channel

(Tanouye *et al.*, 1981; Salkoff, 1983). At the same time, cloning strategies using transposable elements (Bingham *et al.*, 1981; Modolell *et al.*, 1983; Searles *et al.*, 1982) combined with the recent elucidation of the molecular mechanism for the P–M hybrid dysgenesis (Rubin *et al.*, 1982) suggest an obvious and attractive approach to the cloning of *Shaker* gene(s): If a transposable p element can be inserted into the *Shaker* locus thereby causing a mutation of the *Shaker* gene(s), the previously cloned p element can be used to pull out DNA sequences adjacent to the p element insert in the *Shaker* locus.

In addition to providing a starting point for cloning, dysgenesis-induced *Shaker* mutants will also supply abundant new mutations that would be useful in later molecular analysis. Identification of *Shaker* gene(s) within the expected 100 kb of DNA sequences cloned from the *Shaker* locus will rely heavily on molecular analysis of mutations and rearrangements in the *Shaker* locus. By alignment of the genetic map with the physical map of cloned DNA, we will be able to define the sequences encoding particular functions. Since several mutants appear to have a p element inserted in the region of the *Shaker* locus, the location of insertion can be mapped by comparison of the restriction fragments of DNA from wild-type and from *Shaker* mutants. This will allow us to correlate specific phenotypes with specific DNA sequences: The different mutations have been found to show a wide spectrum of *Shaker* phenotypes. Some cause the extreme null phenotype in both nerves and muscles whereas others affect nerves much more than muscles. Should there be multiple structural genes (defined in a molecular analysis as sequences that could encode proteins) within the cloned region, molecular analyses of *Shaker* mutations will be crucial to the assignment of function to different coding sequences. Similarly, should there be different DNA sequences that are not expressed but play a regulatory role, controlling for instance the tissue specificity of expression of nearby sequences, they may be identified in these analyses as well.

There are two indications that the *Shaker* locus may be genetically complex. Rearrangement breakpoints covering a substantial region around 16F give some evidence of a shaking phenotype (Tanouye *et al.*, 1981). Additionally, there is more than one phenotypic consequence of mutations at the *Shaker* locus. Are these different alleles of a complicated gene or are there multiple structural genes with related phenotypes encoded in the region? Because a number of *Shaker* mutations are not completely recessive, complementation studies are unable to resolve the issue. Since genetic recombination experiments thus far have not revealed any widely separated *Shaker* mutations, we believe it will be possible to clone the entire *Shaker* region and define its detailed genetic make up by correlating coding regions with the molecularly mapped positions of numerous *Shaker* mutations of known phenotypes.

How does one demonstrate that the *Shaker* locus contains the structural genes for A channel? If the A channel were composed of a single subunit, one might expect an expression system such as the *Xenopus* oocytes to be useful, as has been the case for acetylcholine receptors (Barnard *et al.*, 1982). In this case, one would need to use cloned *Shaker* DNA to isolate the appropriate messenger RNA so that the latter may be injected into cells like *Xenopus* oocytes. If the A channel were composed of more than one subunit and not all subunits were encoded by DNA from the *Shaker* locus, one might first identify the DNA sequences in the *Shaker* locus that codes for one of the subunits. This might be done by identifying sequences that could hybridize to and remove a messenger RNA that is necessary for the formation of functional A channels. DNA transformation of *Drosophila* provides another powerful method that could be used to show that a particular DNA sequence includes the information damaged by *Shaker* mutations. In this approach, *Drosophila* embryos whose *Shaker* locus had been deleted (e.g., $B55^D/W32^P$) would be injected with different DNA segments from the *Shaker* locus that were ligated to specifically constructed fragments of the transposable p element designed to increase the frequency of integration, stable maintenance, and correct expression of the injected sequences (Rubin and Spradling, 1982; Scholnick *et al.*, 1983; Spradling and Rubin, 1983; Goldberg *et al.*, 1983). The DNA segment that restores A channel in the transformed fly is likely to contain the structural gene and may be used to generate large amounts of its gene products. By using specialized cloning vectors that will allow expression of the *Shaker* DNA that encodes a subunit of the A channel, one may then generate antibodies to the protein product and use them in the purification and molecular characterization of A channel polypeptides. Analysis of the molecular defects in certain mutants, such as Sh^5, that produce altered channels should help define functional domains in the channels. These analyses might also be extended to determine whether different K^+ channels in *Drosophila* share structural homology and whether their evolutionary conservation permits the identification of other channel genes in other organisms.

ACKNOWLEDGMENTS

We are grategul to Dr. W. Engels for providing the P strain *Drosophila* stocks, Dr. G. M. Rubin for providing the p element-containing plasmids, and Ms. Louise Evans for doing chromosomal squashes and the art work. This study was supported by NIH Grant R01 NS15963 to L. Y. Jan and NIH postdoctoral fellowships to C. Laffer, L. Salkoff, and L. Timpe.

REFERENCES

Adams, D. J., Smith, S. J., and Thompson, S. H. (1980). Ionic currents in molluscan soma. *Annu. Rev. Neurosci.* **3,** 141.

Adams, P. R., Brown, D. A., and Constanti, A. (1982). Pharmacological inhibition of the M-current. *J. Physiol. (London)* **332,** 223.

Barnard, E. A., Mileda, R., and Sumikawa, K. (1982). Translation of exogenous messenger RNA encoding for nicotinic acetylcholine receptors produces functional receptors in *Xenopus* oocytes. *Proc. R. Soc. London Ser. B* **215,** 241.

Benton, W. D., and Davis, R. W. (1977). Screening γgt recombinant clones by hybridization to single plaques *in situ*. *Science* **196,** 18O.

Bingham, P. M., Levis, R. and Rubin, G. M. (1981). Cloning of DNA sequences from the *white* locus of *D. melanogaster* by a novel and general method. *Cell* **25,** 693.

Byrne, J. H. (1980). Analysis of ionic conductance mechanisms in motor cells mediating inking behavior in *Aplysia californica*. *J. Neurophysiol.* **43,** 630.

DePeyer, J. E., Cachelin, A. B., Levitan, I. B., and Reuter, H. (1982). Ca^{2+}-activated K^{+} conductance in internally perfused neurons is enhanced by protein phosphorylation. *Proc. Natl. Acad. Sci. U.S.A.* **79,** 4207.

Engels, W. R. (1981). Hybrid dysgenesis in *Drosophila* and the stochastic loss hypothesis. *Cold Spring Harbor. Symp. Quant. Biol.* **45,** 561.

Goldberg, D. A. Posakony, J. W., and Manialis, T. (1983). Correct developmental expression of a cloned alcohol dehydrogenase gene transduced into the Drosophila germ line. *Cell* **34,** 59–73.

Hugues, M., Duval, D., Kitabgi, P., Lazkunski, M., and Vincent, J. P. (1982). Preparation of a pure monoiodo derivative of the bee venom neurotoxin apamin and its binding properties to rat brain synaptosomes. *J. Biol. Chem.* **257,** 2762.

Jan, L. Y., Barbel, S., Timpe, L., Laffer, C., Salkoff, L., O'Farrell, P., and Jan, Y. N. (1983). Mutating a gene for a K^{+} channel by hybrid dysgenesis: an approach to the cloning of the *Shaker* locus in *Drosophila*. *Cold Spring Harbor Symp. Quant. Biol.*

Jan, Y. N., and Jan, L. Y. (1980). Genetic dissection of synaptic transmission in *Drosophila melanogaster*. *In* "Insect Neurobiology and Pesticide Action" (F. E. Rickett, ed.), pp. 161–168. Society of Chemical Industry, London.

Jan, Y. N., Jan, L. Y., and Dennis, M. J. (1977). Two mutations of synaptic transmission in *Drosophila*. *Proc. R. Soc. London Ser B* **198,** 87.

Klein, M., and Kandel, E. R. (1980). Mechanism of calcium current modulation underlying presynaptic facilitation and behavioral sensitization in *Aplysia*. *Proc. Natl. Acad. Sci. U.S.A.* **77,** 6912.

Klein, M., Camardo, J., and Kandel, E. R. (1982). Serotonin modulates a specific potassium current in the sensory neurons that show presynaptic facilitation in *Aplysia*. *Proc. Natl. Acad. Sci. U.S.A.* **79,** 5713.

Llinas, R., Sugimori, M., and Simon, S. M. (1982). Transmission by presynaptic spike-like depolarization in the squid giant synapse. *Proc. Natl. Acad. Sci. U.S.A.* **79,** 2415.

Maniatic, T., Hardison, R. C., Lacy, E., Lauer, J., O'Connell, C., Quon, D., Sim, D. K., and Efstratiadis, A. (1978). The isolation of structural genes from libraries of eucaryotic DNA. *Cell* **15,** 687.

Modolell, J., Bender, W., and Meselson, M. (1983). *Drosophila melanogaster* mutations suppressible by the suppressor of *Hairy-wing* are insertions of a 7.3-kilobase mobile element. *Proc. Natl. Acad. Sci. U.S.A.* **80,** 1678.

Pardue, M. L., and Gall, J. G. (1975). Nucleic acid hybridization to the DNA of cytological preparations. *Methods Cell Biol.* **10,** 1.

Rubin, G. M., and Spradling, A. C. (1982). Genetic transformation of *Drosophila* with transposable element vectors. *Science* **218,** 348.

Rubin, G. M., Kidwell, M. G., and Bingham, P. M. (1982). The molecular basis of P-M hybrid dysgenesis: The nature of induced mutations. *Cell* **29,** 987.

Salkoff, L. (1983). Genetic and voltage-clamp analysis of a *Drosophila* K^{+} channel. *Cold Spring Harbor Symp. Quant. Biol.*

Salkoff, L., and Wyman, R. (1981). Genetic modification of potassium channels in *Drosophila Shaker* mutants. *Nature (London)* **293,** 228.

Scholnick, S. B., Morgan, B. A., and Hirsh, J. (1983). The cloned dopa decarboxylase gene is developmentally regulated when reintegrated into the *Drosophila* genome. *Cell* **34,** 37–45.

Searles, L. L., Jokerst, R. S., Bingham, P. M., Voelker, R. A., and Greenleaf, A. L. (1982). Molecular cloning of sequences from a *Drosophila* RNA polymerase II locus by a p element transposon tagging. *Cell* **31,** 585.

Siegelbaum, S. A., Camardo, J. S., and Kandel, E. R. (1982). Serotonin and cyclic AMP close single K^+ channels in *Aplysia* sensory neurons. *Nature* (*London*) **299,** 413.

Spradling, A. C., and Rubin, G. M. (1983). The effect of chromosomal position on the expression of the Drosophila xanthine dehydrogenase gene. *Cell* **34,** 47–57.

Tanouye, M. A., Ferrus, A., and Fujita, S. C. (1981). Abnormal action potentials associated with the *Shaker* locus of *Drosophila. Proc. Natl. Acad. Sci. U.S.A.* **78,** 6548.

Chapter 5

Sodium Channels in Neural Cells: Molecular Properties and Analysis of Mutants

WILLIAM A. CATTERALL, TOHRU GONOI, AND MARIA COSTA[1]

Department of Pharmacology
University of Washington
Seattle, Washington

I. INTRODUCTION

Electrical excitability is among the most important and characteristic properties of neurons. Most vertebrate cells, including neurons, maintain large ionic gradients across their surface membranes such that the intracellular fluid contains

[1]Present address: Department of Hematology, University of Rochester, School of Medicine and Dentistry, Rochester, New York 14642.

ISBN 0-12-153323-9

a low concentration of Na^+ and Ca^{2+} and a high concentration of K^+ relative to the extracellular fluid. These large ion gradients are maintained by the action of energy-dependent ion pumps specific for Na^+ and K^+ or Ca^{2+}. In addition, essentially all vertebrate cells maintain an internally negative membrane potential of the order of -60 mV since their surface membranes are specifically permeable to K^+, allowing K^+ to leak out of cells faster than Na^+ and Ca^{2+} can leak in. Nerve cells are electrically excitable because of the presence of voltage-sensitive ion channels in their surface membranes that are selective for Na^+, K^+, or Ca^{2+}. One class of Na^+ channels and multiple classes of Ca^{2+} and K^+ channels have been described in neurons. These channels open and close as a function of membrane voltage allowing rapid movement of the appropriate ions down their concentration gradient carrying ionic current into or out of the cell and thereby depolarizing or hyperpolarizing the membrane potential. Depolarization of the cell membrane beyond a threshold value elicits one or a series of conducted action potentials which are initiated in the cell soma or the initial segment of the axon and are conducted down the axon to the nerve terminal. The action potential invades the nerve terminal causing depolarization, release of neurotransmitter into the synaptic cleft, and excitation of succeeding neurons in the pathway or of effector cells such as skeletal muscle. While Na^+, K^+, and Ca^{2+} channels each contribute in an essential way to signal processing and transmission in neurons, the role, mechanism of action, and molecular properties of the voltage-sensitive sodium channel are understood most completely. This article will briefly review the physiological properties of sodium channels, consider recent experiments that have begun to define the nature of the membrane macromolecules that comprise the sodium channel in neurons, and describe a somatic cell genetic approach to analysis of sodium channel properties.

The ionic mechanisms underlying electrical excitability have been defined using the method of voltage clamp (Hodgkin and Huxley, 1952). In this approach the voltage across the excitable membrane is controlled by use of a feedback amplifier circuit, and the ionic currents moving across the membrane in response to step changes in the membrane potential imposed by the experimenter are measured. Experiments using the voltage clamp technique have shown that the initial rapid depolarization during an action potential in nerve axons results from rapid voltage-dependent increases in membrane permeability to sodium ions (Hodgkin and Huxley, 1952). Many different lines of evidence indicate that a selective transmembrane sodium channel is responsible for the rapid sodium permeability increase during the action potential. Selective ion permeation is mediated by a hydrophilic pore containing a sodium-selective ion coordination site designated the ion selectivity filter (Hille, 1971, 1972). Ion conductance through the sodium channel is regulated or "gated" by two separate processes: *activation*, which controls the rate and voltage dependence of opening of the sodium channel following depolarization, and *inactivation,* which controls the rate and voltage dependence of the subsequent closing of the sodium channel during a maintained depolarization (Hodgkin and Huxley, 1952). Estimates of the rate of sodium movement through an activated sodium channel derived from analysis of membrane current noise (Conti *et al.*, 1976; Sigworth, 1980) or recordings of individual sodium channel currents (Sigworth and Neher, 1980) range from 8 to 18 pS corresponding to greater than 10^7 ions/sec/channel at physiological temperature and Na^+ concentration. These rates approach those for diffusion through free solution and imply that the residence time of an individual sodium ion in the channel is short and the interactions with the channel weak.

Analysis of sodium channel properties by the voltage clamp method has provided a detailed description of the three essential functional properties of sodium channels: voltage-dependent activation, inactivation, and selective ion transport. However, an understanding of the molecular basis of neuronal excitability requires identification of the membrane macromolecules that comprise the ionic channels, solubilization and purification of these channel components, and correlation of their structural features with the known functional properties of sodium channels. New approaches have been developed to attack these problems.

II. NEUROTOXINS AS MOLECULAR PROBES OF SODIUM CHANNELS

Neurotoxins which bind with high affinity and specificity to voltage-sensitive sodium channels and modify their properties have provided the essential tools for identification and purification of sodium channels. Four different groups of neurotoxins which act at four different neurotoxin receptor sites on the sodium channel have been useful in these studies (Table I). Neurotoxin receptor site 1 binds the water-soluble, heterocyclic guanidines tetrodotoxin (TTX) and saxitox-

TABLE I
NEUROTOXIN RECEPTOR SITES ON THE SODIUM CHANNEL

Site	Neurotoxins	Physiological effect
1	Tetrodotoxin, saxitoxin	Inhibit ion transport
2	Veratridine, batrachotoxin, grayanotoxin, aconitine	Cause persistent activation
3	North African α-scorpion toxins, sea anemone toxins	Slow inactivation
4	American β-scorpion toxins	Enhance activation

in (STX). These toxins inhibit sodium channel ion transport by binding to a common receptor site which is thought to be located near the extracellular opening of the ion-conducting pore of the sodium channel (Narahashi, 1974; Ritchie and Rogart, 1977; Catterall, 1980).

Neurotoxin receptor site 2 binds several lipid-soluble toxins including grayanotoxin and the alkaloids veratridine, aconitine, and batrachotoxin (Catterall, 1980; Albuquerque and Daly, 1976). The competitive interactions among these four toxins at neurotoxin receptor site 2 have been confirmed by direct measurements of specific binding of [^{3}H]batrachotoxinin A 20α-benzoate to sodium channels (Brown *et al.*, 1981; Catterall *et al.*, 1981). These toxins cause persistent activation of sodium channels at the resting membrane potential by blocking sodium channel inactivation and shifting the voltage dependence of sodium channel activation to more negative membrane potentials (Catterall, 1980). Therefore, neurotoxin receptor site 2 is likely to be localized on a region of the sodium channel involved in voltage-dependent activation and inactivation.

Neurotoxin receptor site 3 binds polypeptide toxins purified from North African scorpion venoms or sea anemone nematocysts. These toxins slow or block sodium channel inactivation. They also markedly enhance persistent activation of sodium channels by the lipid-soluble toxins acting at neurotoxin receptor site 2 (reviewed in Catterall, 1980). The affinity for binding of ^{125}I-labeled derivatives of the polypeptide toxins to neurotoxin receptor site 3 is reduced by depolarization. The voltage dependence of scorpion toxin binding is closely correlated with the voltage dependence of sodium channel activation (Catterall, 1979). These experiments indicate that neurotoxin receptor site 3 is located on part of the sodium channel that undergoes a conformational change during voltage-dependent channel activation leading to markedly reduced affinity for scorpion toxin. Therefore, scorpion toxin and sea anemone toxin bind to voltage-sensing or gating structures of sodium channels.

Neurotoxin receptor site 4 binds a new class of scorpion toxins that has also proven valuable in studies of sodium channels. Cahalan (1975) showed that the venom of the American scorpion *Centruroides sculpturatus* modifies sodium channel activation rather than inactivation. Pure toxins from several American scorpions have a similar action (Wang and Strichartz, 1982; Meves *et al.*, 1982; Couraud *et al.*, 1982). These toxins bind to a new receptor site on the sodium channel (Jover *et al.*, 1980; Barhanin *et al.*, 1982a) and have therefore been designated β-scorpion toxins.

These several neurotoxins provide specific high-affinity probes for distinct regions of the sodium channel structure. They have been used to detect and localize sodium channels in neuronal cells, as well as to identify and purify the protein components of sodium channels that bind these toxins and to analyze their structural and functional properties.

III. STRUCTURE OF THE SODIUM CHANNEL

A. Identification of Protein Components of Sodium Channels in Neuronal Membranes

Measurements of the distribution and density of sodium channels indicate that, with the exception of the very small amount of specialized membrane at the node of Ranvier, sodium channels are a minor component of excitable membranes. These results emphasize the need for highly specific probes to identify the macromolecules that comprise the sodium channel. The neurotoxins that bind to sodium channels with high affinity and specificity have provided the tools needed in such experiments. Direct chemical identification of sodium channel components *in situ* was first achieved by specific covalent labeling of neurotoxin receptor site 3 with a photoreactive azidonitrobenzoyl derivative of the α-scorpion toxin from *Leiurus quinquestriatus*. The photoreactive toxin derivative is allowed to bind specifically to sodium channels in the dark. Irradiation with UV light then chemically activates the arylazide group, which covalently reacts with the scorpion toxin receptor site on the sodium channel. Analysis of covalently labeled synaptosomes by polyacrylamide gel electrophoresis under denaturing conditions in sodium dodecyl sulfate (SDS) to separate synaptosomal proteins by size reveals specific covalent labeling of two polypeptides, which have subsequently been designated the α and β1 subunits of the sodium channel (Beneski and Catterall, 1980). These proteins, as assessed by polyacrylamide gel electrophoresis in SDS, have molecular weights of 270,000 and 39,000, respectively (Hartshorne and Catterall, 1981; Hartshorne *et al.*, 1982).

The covalent labeling of these two polypeptides in synaptosomes was shown to be specific by competitive inhibition with excess unlabeled scorpion toxin or by blockage of voltage-dependent binding of scorpion toxin by membrane depolarization (Beneski and Catterall, 1980). The α subunit of the sodium channel could also be covalently labeled with azidonitrobenzoyl scorpion toxin in electrically excitable neuroblastoma cells (Beneski and Catterall, 1980).

The β-scorpion toxins derived from American scorpion venoms have also been used to label neurotoxin receptor site 4 on the sodium channel (Barhanin *et al.*, 1982b). Toxin γ from *Tityus serrulatus* was covalently attached to its receptor site by covalent cross-linking with disuccinimidyl suberate. A single 270,000-Da polypeptide was labeled in rat brain synaptosomes. Thus, the receptor site for the β-scorpion toxins is located on or near the α subunit of the sodium channel as previously found for the α-scorpion toxins acting at neurotoxin receptor site 3.

B. The Molecular Size of the Sodium Channel

The first indications of the molecular size of the neuronal sodium channel *in situ* were derived from radiation inactivation studies (Levinson and Ellory, 1973). In these experiments, membrane preparations from pig brain were irradiated with X rays, and the decrease in the number of functional tetrodotoxin binding sites was measured as a function of radiation dose. From these data, the size of the membrane target can be determined since larger targets are more likely to be hit and are therefore inactivated at a lower radiation dose. Applying target theory, Levinson and Ellory concluded that a 230,000-Da structure was required for toxin binding. These experiments have been repeated by Barhanin *et al.* (1982b), who compared the target size of the sodium channel assessed by either tetrodotoxin binding or *Tityus serrulatus* toxin binding. In each case, the target size, approximately 270,000 Da, was in reasonable agreement with the earlier work. This size estimate might correspond to the molecular weight of the entire sodium channel or to the molecular weight of a protein subunit that is essential for binding these neurotoxins.

The molecular size of the intact sodium channel protein has been measured by hydrodynamic studies of the detergent-solubilized channel. The STX- and TTX-binding component of sodium channels was first solubilized with retention of high affinity and high specificity of toxin binding from garfish olfactory nerve membrane using nonionic detergents (Benzer and Raftery, 1973; Henderson and Wang, 1972). Similar techniques have now been applied to sodium channels in mammalian brain (Krueger *et al.*, 1979; Catterall *et al.*, 1979). In contrast to the ease of solubilization of the sodium channel with retention of STX- and TTX-binding activity at neurotoxin receptor site 1, both neurotoxin receptor site 2 (unpublished observations) and neurotoxin receptor site 3 (Catterall *et al.*, 1979) lose high-affinity neurotoxin-binding activity on solubilization. The molecular weight of the solubilized sodium channel from rat brain has been estimated by hydrodynamic studies to be 601,000 (Hartshorne *et al.*, 1980). Since the detergent–channel complex contains 0.9 g Triton X-100 and phosphatidylcholine per gram of protein, the molecular weight of the sodium channel protein solubilized from rat brain is 316,000 (Hartshorne *et al.*, 1980). This represents the size of the entire sodium channel as solubilized in detergents and corresponds to a complex of three nonidentical protein subunits as described below. If the channel protein is spherical in shape, the diameter indicated by these results is 118 Å. Thus, the channel protein is much larger than the postulated transmembrane pore through which Na^+ moves, which is proposed to be 3×5 Å at its narrowest point—the ion selectivity filter (Hille, 1972).

C. Protein Subunits of the Purified Sodium Channel from Mammalian Brain

The ability to solubilize the sodium channel from brain membranes in a well-defined monomeric form with retention of binding activity for saxitoxin and tetrodotoxin has allowed purification by a sequence of conventional protein separation procedures (Hartshorne and Catterall, 1981). The current purification scheme developed in this laboratory employs anion exchange chromatography on DEAE-Sephadex, adsorption chromatography on hydroxyapatite gel, affinity chromatography on wheat germ agglutinin attached to Sepharose 4B, and, finally, velocity sedimentation through sucrose gradients (Hartshorne *et al.*, 1982; Hartshorne and Catterall, 1984). The purified sodium channel preparation binds 0.9 mol of saxitoxin per mol of sodium channel of 316,000 Da. Assuming that the sodium channel binds only one saxitoxin molecule, these data indicate that at least 90% of the protein in the purified preparation must be associated with the sodium channel.

The protein subunits of the sodium channel have been analyzed by denaturation of the protein at 100°C in the presence of SDS and β-mercaptoethanol followed by separation according to molecular weight by electrophoresis in polyacrylamide gels. Two protein bands with molecular weights of 260,000 and 38,000 are resolved from the purified sodium channel with this technique (Fig. 1;

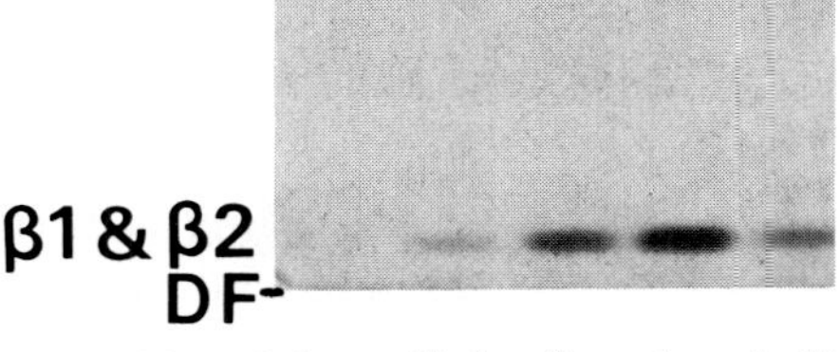

FIG. 1. Polypeptide composition of the purified sodium channel. The peak fraction from a sucrose gradient used as the final purification step of the sodium channel was incubated at 100°C in sodium dodecyl sulfate and β-mercaptoethanol and subjected to electrophoresis in a 5.5 to 12% gradient polyacrylamide gel in the presence of sodium dodecyl sulfate as previously described (Hartshorne *et al.*, 1982). The polypeptide components of the sodium channel were visualized by a sensitive silver staining procedure. The origin (Or), dye front (DF), and the subunits of the sodium channel are labeled.

TABLE II
SUBUNIT COMPOSITION OF THE SODIUM CHANNEL PURIFIED FROM RAT BRAIN

	Molecular weight	Probable stoichiometry
Native sodium channel	316,000	
α Subunit	260,000	1.0
β1 Subunit	39,000	1.0
β2 Subunit	37,000	1.0

Hartshorne *et al.*, 1982; Hartshorne and Catterall, 1984). These two bands, which we designate α and β, contain greater than 90% of the protein in the purified preparation and co-migrate precisely with the saxitoxin-binding activity of the sodium channel during velocity sedimentation on sucrose gradients (Hartshorne and Catterall, 1984). Thus, the sodium channel isolated from rat brain appears 90% pure by both chemical and functional criteria and consists of at least two classes of subunits.

The β protein band illustrated in Fig. 1 actually contains two nonidentical polypeptides (Table II) with molecular weights of 39,000 (β1) and 37,000 (β2, Hartshorne *et al.*, 1982). These two polypeptides are distinguished by two properties: β2 is covalently attached to the α subunit by disulfide bonds, whereas β1 is noncovalently bound. The β1 chain is covalently labeled by photoreactive scorpion toxin derivatives whereas β2 is not. These results show that the sodium channel as isolated from rat brain is a complex of three nonidentical subunits. Comparison of the molecular weights of the individual subunits with the molecular weight of the entire complex (Table II) suggests that the subunits are present in a 1 : 1 : 1 stoichiometry.

D. Reconstitution of Sodium Channel Function from Purified Components

The purified sodium channel from rat brain binds [^{3}H]saxitoxin and tetrodotoxin with the same affinity as the native sodium channel and therefore contains neurotoxin receptor site 1 of the sodium channel in an active form (Hartshorne and Catterall, 1984). The purified channel also contains the α and β1 subunits identified as components of neurotoxin receptor site 3 by photoaffinity labeling with scorpion toxin (Beneski and Catterall, 1980), although after solubilization the binding activity for scorpion toxin is lost (Catterall *et al.*, 1979). However, the purified channel does not have binding activity for neurotoxins at receptor site 2 and cannot transport sodium in the detergent-solubilized state. Reconstitution of these sodium channel functions from purified components is the only rigorous proof that the proteins identified and purified on the basis of their neurotoxin-binding activity are indeed sufficient to form a functional voltage-sensitive ion channel. In addition, successful reconstitution will provide a valuable experimental preparation for biochemical analysis of the structure and function of sodium channels.

Several groups have successfully restored aspects of sodium channel function from detergent-solubilized brain membranes, thus showing that detergent solubilization does not irreversibly destroy channel function (Villegas and Villegas, 1981; Malysheva *et al.*, 1980; Tamkun and Catterall, 1981; Goldin *et al.*, 1980). More recently, sodium channel ion transport has been successfully reconstituted from sodium channels substantially purified from rat brain and skeletal muscle (Weigele and Barchi, 1982; Talvenheimo *et al.*, 1982). We have now applied these methods to essentially homogeneous preparations of sodium channels from rat brain (Tamkun *et al.*, 1984). Purified sodium channels in Triton X-100 solution are supplemented with phosphatidylcholine dispersed in Triton X-100 and the detergent is removed by adsorption to polystyrene beads. As the detergent is removed, phosphatidylcholine vesicles with a mean diameter of 1800 Å are formed containing an average of 0.75 to 2 sodium channels per vesicle. The functional activities of the sodium channel can then be assessed in neurotoxin-binding and ion flux experiments.

Figure 2A illustrates the time course of $^{22}Na^+$ influx into phosphatidylcholine vesicles containing purified sodium channels. In these experiments, the vesicle preparation was incubated for 2 min with veratridine to activate sodium channels and then diluted into medium containing $^{22}Na^+$ to initiate influx into vesicles. After time intervals of 10 sec to 15 min, the samples were applied to columns of a cation exchange resin and washed through with isotopic sucrose. The vesicles and intravesicular $^{22}Na^+$ pass through the column while extravesicular $^{22}Na^+$ is quantitatively bound to the column. $^{22}Na^+$ influx is then measured by γ counting. Influx into vesicles under control conditions is slow (Fig. 2A). Incubation with veratridine increases the initial rate of influx 10- to 15-fold. When tetrodotoxin is present in both the intravesicular and extravesicular phases, the veratridine-dependent increase in initial rate of $^{22}Na^+$ influx is blocked nearly completely (Fig. 2A). Half-maximal activation is observed with 28 μM veratridine and half-maximal inhibition with 14 nM tetrodotoxin in close agreement with $K_{0.5}$ values for the action of these toxins on native sodium channels. These results show that the purified sodium channel regains the ability to mediate neurotoxin-stimulated ion flux after incorporation into phosphatidylcholine vesicles. Evidently, the purified channel retains neurotoxin receptor site 2 and the ion-conducting pore of the sodium channel.

Ion transport by neurotoxin-activated sodium channels in neural membranes is selective, although the rate of transport of large cations such as Rb^+ and Cs^+ relative to the rate of transport of Na^+ is significantly greater than when chan-

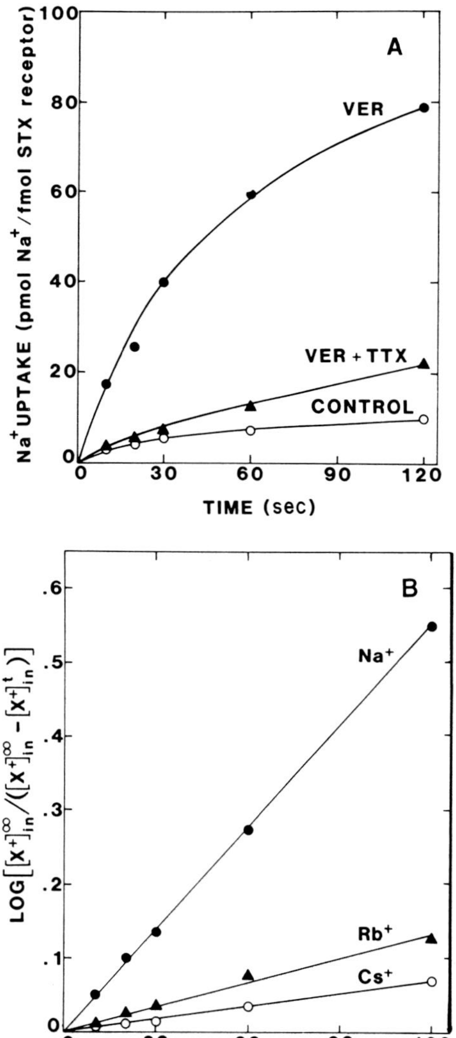

FIG. 2. Neurotoxin-activated ion flux mediated by the purified and reconstituted sodium channel. Sodium channels were purified to near homogeneity and incorporated into phosphatidylcholine vesicles as described previously (Tamkun *et al.*, 1984). (A) The vesicles were incubated for 2 min at 36°C with no additions (○) or 100 μM veratridine (●) and then diluted into medium containing 10 μCi/ml ^{22}NaCl and the same neurotoxins to give a final external concentration of 13 mM NaCl. The initial rate of $^{22}Na^{+}$ influx was measured for the indicated times by separating the vesicles containing $^{22}Na^{+}$ from the medium by passage over a column of Dowex 50W-X8 (Tris form) and elution with isotonic sucrose. Transport of $^{22}Na^{+}$ into the vesicles was measured by gamma counting. An identical vesicle sample (▲) containing 1 μM tetrodotoxin within the vesicles was incubated for 2

nels are activated by membrane depolarization (Khodorov, 1978; Frelin *et al.*, 1981; Huang *et al.*, 1979). Figure 2B illustrates the initial rate of influx of $^{22}Na^+$, $^{86}Rb^+$, and $^{137}Cs^+$ through veratridine-activated sodium channels in reconstituted vesicles. The purified and reconstituted sodium channel retains ion selectivity with permeability ratios of 0.25 for Rb^+ and 0.12 for Cs^+ relative to Na^+. These permeability ratios compare favorably to those of native sodium channels activated by neurotoxins.

While sodium channels reconstituted into phosphatidylcholine vesicles can transport sodium, these channels do not bind α-scorpion toxin at neurotoxin receptor site 3. In contrast, if purified sodium channels are incorporated into vesicles composed of a mixture of phosphatidylcholine and brain lipids, scorpion toxin binding is recovered. The toxin-binding reaction is of high affinity (K_D = 43 n*M*) and a mean of 0.76 ± 0.08 mol scorpion toxin is bound per mol of purified sodium channel (Tamkun *et al.*, 1984). The brain lipid fraction was prepared by $CHCl_3/Ch_3OH$ extraction followed by silicic acid chromatography, and was found to be protein free by gel electrophoresis and sensitive silver staining. Therefore, we conclude that brain lipids are essential to restore the scorpion toxin receptor site to the same functional state as in native membranes. Since the affinity for scorpion toxin binding to synaptosomal sodium channels is dependent upon the functional state of the sodium channel as reflected in the voltage dependence of toxin binding and its allosteric modulation by alkaloid toxins such as veratridine (Catterall, 1979; Krueger and Blaustein, 1980; Tamkun and Catterall, 1981; Jover *et al.*, 1978; Ray *et al.*, 1978), components of the brain lipid mixture may also be required for other functional activities of the channel. Since the scorpion toxin binding in reconstituted vesicles containing brain lipid is not voltage dependent (Tamkun *et al.*, 1984), the lipid environment provided by the mixture of phosphatidylcholine and whole brain lipid may not be optimal for channel function. Further analysis of the lipid environment necessary for recovery of channel function may reveal requirements for specific membrane lipids for voltage sensitivity.

The results of these reconstitution experiments show that the purified sodium channel preparation from rat brain consisting of a stoichiometric complex of the α, β1, and β2 subunits is sufficient to mediate most of the functions of the sodium channel that can be measured using biochemical methods. These include neurotoxin binding and action at neurotoxin receptor sites 1 through 3 and selective neurotoxin activated ion flux. However, in excitable membranes so-

---

min at 36°C with 100 μ*M* veratridine and 1 μ*M* tetrodotoxin, and $^{22}Na^+$ influx was measured in the presence of the same neurotoxins as described above. (B) Similar experiments to those in (A) were carried out in 10 μCi/ml ^{22}NaCl, 10 μCi/ml ^{86}RbCl, and 10 μCi/ml ^{137}CsCl. Influx of $^{22}Na^+$ under control conditions has been subtracted. The data are plotted as fractional equilibration of the intravesicular volume on semilogarithmic axes.

dium channels are normally activated and inactivated by changes in membrane potential. Purified and reconstituted channels have not yet been tested for their ability to activate and inactivate on depolarization. Assessment of this important aspect of channel function awaits application of methods to record ionic currents mediated by purified channels in large reconstituted vesicles or planar bilayers. These experiments will provide the final test for the functional integrity of the purified and reconstituted channels.

IV. ANALYSIS OF NEUROBLASTOMA CELLS WITH MISSING OR ALTERED SODIUM CHANNELS

Somatic cell genetics has provided an alterantive approach to analysis of complex biochemical and physiological processes in mammalian cells. The successful identification and resolution of the components of the adenylate cyclase system provide an excellent example (Bourne *et al.*, 1975; Ross *et al.*, 1978). This approach requires development of a selection method to isolate variants in the process of interest and analytical techniques to determine the properties of the variants. As in the biochemical experiments described above, studies of sodium channels by somatic cell genetic methods have relied upon specific neurotoxins that modify channel properties.

A. Selection of Neurotoxin-Resistant Neuroblastoma Cell Lines

Inhibition of ion pumping in cells by ouabain is cytotoxic and has been used successfully to select ouabain-resistant cell lines (Baker *et al.*, 1974). This cytotoxicity probably results from alteration of the intracellular ion concentrations. Because neurotoxins that cause persistent activation of sodium channels in cells greatly increase sodium permeability and intracellular sodium concentration, these toxins were tested as selective agents for neuroblastoma cell clones with missing or altered sodium channels (West and Catterall, 1979). The effect of growth of neuroblastoma cells (clone N18) in the presence of neurotoxins upon plating efficiency is illustrated in Table III. Growth in the presence of 40 μM veratridine plus 50 nM scorpion toxin is highly cytotoxic, reducing plating efficiency to 0.05% of control (Table III). Inclusion of 1 μM tetrodotoxin in the growth medium blocks the persistently activated sodium channels and reverses the cytotoxic effect of veratridine plus scorpion toxin (Table III). These results show that persistent activation of sodium channels is cytotoxic.

To isolate neuroblastoma cell clones resistant to these neurotoxins, cells were plated in selective medium containing 40 μM veratridine plus 50 nM scorpion

TABLE III
EFFECT OF NEUROTOXINS ON PLATING EFFICIENCY OF NEUROBLASTOMA CELLS

Additions	Plating efficiency (%)
None	100
40 μ*M* veratridine, 50 n*M* scorpion toxin	0.05
40 μ*M* veratridine, 50 n*M* scorpion toxin, 1 μ*M* tetrodotoxin	86

toxin and grown for 2 weeks. Resistant colonies arising by that time were encircled with porcelain cylinders and cells were suspended and replated at low density. Single-cell clones were then isolated by encircling single cells with porcelain cylinders and allowing them to grow to give full-sized colonies. Single-cell clones were expanded into larger culture vessels and stocks were prepared and stored for biochemical and electrophysiological analysis. Single-cell clones isolated in this manner had plating efficiencies in selective growth medium ranging from 11 to 71% of those in normal medium compared to values of 0.02–0.05% for the N18 cell line (Table IV). The resistance to neurotoxins was retained during 100 generations of growth in nonselective medium (West and Catterall, 1979).

B. Phenotypic Properties of Variant Cell Clones

Neurotoxin-resistant cell clones were initially analyzed by ion flux procedures, which measure the changes in sodium permeability of neuroblastoma cells caused by persistent activation of sodium channels with 40 μ*M* veratridine plus 200 n*M* scorpion toxin. These procedures indicated that the resistant colonies fall into three classes (Table IV). *Sodium channel-deficient* cell lines were the most prevalent phenotype, comprising approximately 80% of all resistant clones. These cell lines have less than 10% of the neurotoxin-stimulated $^{22}Na^+$ influx of N18 cells. They also have less than 10% as many high-affinity scorpion toxin receptor sites (West and Catterall, 1979). These results are consistent with the conclusion that *sodium channel-deficient* cell lines have a sharply reduced number of functional sodium channels.

The second phenotypic class of neurotoxin-resistant cells lines is *scorpion toxin resistant*. It comprises approximately 10% of resistant clones. The two cell lines we have analyzed extensively have reduced levels of neurotoxin-activated

TABLE IV
PHENOTYPES OF NEUROTOXIN-RESISTANT NEUROBLASTOMA CELLS

Phenotype	Number of clones analyzed	Plating efficiency (%)	Neurotoxin-activated $^{22}Na^+$ influx (nmol/min/mg)	K_D for scorpion toxin (nM)
N18		0.02	116	0.7
Sodium channel deficient	10	11–71	3–12	—
Scorpion toxin resistant	2	0.6, 36	32, 40	19, 40
Parental type	2	32, 45	35, 38	1.0

$^{22}Na^+$ influx (Table IV) indicating the presence of fewer sodium channels. However, the extent of reduction is not sufficient to account for the resistance to cytotoxicity. In addition to a reduced number of sodium channels, these cells also have reduced affinity for scorpion toxin. This is most clearly demonstrated by measuring the concentration dependence of enhancement of veratridine activation of sodium channels by scorpion toxin (West and Catterall, 1979). In these experiments, neuroblastoma cell cultures are incubated with increasing concentrations of scorpion toxin for 60 min and then challenged with 200 μM veratridine in the presence of $^{22}Na^+$ and ouabain to block the sodium pump. Veratridine activates approximately 8% of sodium channels in N18 cells by itself and approximately 60% in the presence of a saturating concentration of scorpion toxin (Catterall, 1977). Figure 3 illustrates the concentration–effect relationship for this action of scorpion toxin in N18 cells and LV10 cells, a *scorpion toxin-resistant* cell line. The apparent K_D for scorpion toxin is increased approximately 50-fold. A 27-fold increase is observed for clone LV30 (Costa and Catterall, 1982). In both these cells lines, apparent K_D values for tetrodotoxin block of sodium channels at neurotoxin receptor site 1 and batrachotoxin activation of sodium channels at neurotoxin receptor site 2 are unchanged (West and Catterall, 1979; Costa and Catterall, 1982). Evidently, these cells contain sodium channels with a specific alteration in the scorpion toxin receptor site.

The third phenotypic class accounts for approximately 10% of resistant colonies and has sodium channels with parental properties (Table IV; West and Catterall, 1979). We assume that these cells have other inherited alterations that allow them to grow in spite of the markedly increased sodium influx caused by persistent activation of sodium channels. Such alterations might include increased numbers of sodium pumps or decreased sensitivity of essential processes to intracellular ionic environment.

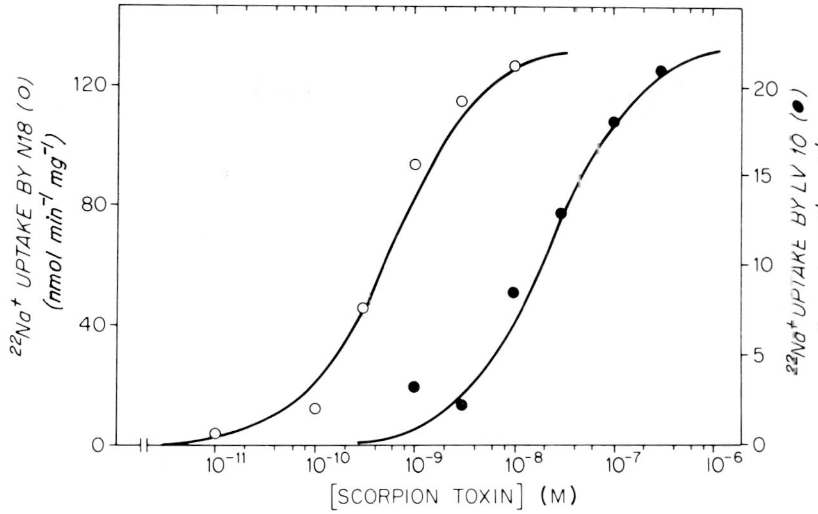

FIG. 3. Concentration dependence of scorpion toxin action on sodium channels in normal and variant neuroblastoma cells. The enhancement of veratridine-stimulated $^{22}Na^+$ influx was measured. N18 cells (○) or LV10 cells (●) were incubated for 30 min at 36°C with the indicated concentrations of scorpion toxin. The initial rate of $^{22}Na^+$ influx was then measured for 30 sec in the presence of 200 μ*M* veratridine as described previously (West and Catterall, 1979). The increment over veratridine-stimulated $^{22}Na^+$ influx is plotted in nmol min^{-1} mg^{-1}. Note the different ordinate axes for N18 and LV10.

C. Effects of Mutagenesis on the Frequency and Phenotype of Resistant Cell Lines

It is of interest to know whether treatment of cells with known mutagens increases the frequency of resistant colonies. We used the potent point mutagen *N*-methyl-*N*′-nitro-*N*-nitrosoguanidine (MNNG) to examine this point. Cells were treated with mutagen for 2 hr and then grown for 7 days to allow expression of altered phenotypes. Selective medium was added and the resistant colonies counted and isolated as independent clones for study. Mutagen treatment increased the frequency of resistant colonies up to 16-fold at concentrations ranging from 0.25 to 2 μg MNNG/ml (Costa and Catterall, 1982). In this same concentration range, MNNG also increased the frequency of colonies resistant to 6-thioguanine. This phenotype is known to arise by mutations in the enzyme hypoxanthineguanine phosphoribosyltransferase (EC 2.4.2.8) (Fenwick and Caskey, 1975). Therefore, our results support the conclusion that neurotoxin-resistant neuroblastoma cells arise by mutational events.

To determine whether any of the three different neurotoxin-resistant phenotypes is selectively increased by mutagenesis, we analyzed the phenotype of 25 independent resistant clones isolated from cell stocks treated with 2 μg/ml MNNG. As in the absence of mutagenesis, approximately 80% of the clones were *sodium channel deficient* (19/25), approximately 10% were *scorpion toxin resistant* (3/25), and approximately 10% were of the parental type with respect to sodium channels (3/25). Since the frequency of all three phenotypes is increased by mutagenesis, we conclude that all phenotypes can arise through mutational events. The nature of such mutations remains uncertain. *Sodium channel-deficient* clones might arise as a result of mutations in regulatory genes controlling neuronal development, sodium channel levels, or biosynthesis and processing of membrane glycoproteins. This phenotype might also arise from structural gene mutations that lead to nonfunctional sodium channels or channels that cannot be properly inserted into membranes. *Scorpion toxin-resistant* clones are the best candidates for mutations in structural genes but these might also arise from mutational alterations in general membrane properties that reduce affinity for scorpion toxin.

D. Voltage Clamp Analysis of Scorpion Toxin-Resistant Neuroblastoma Cells

Leiurus scorpion toxin slows sodium channel inactivation by binding to neurotoxin receptor site 3 on the sodium channel. Its binding is voltage dependent. These characteristics of the action of the toxin suggest the possibility that scorpion toxin-resistant sodium channels might have alterations in normal sodium channel gating. Recently, whole cell patch voltage clamp procedures have been developed that allow accurate and convenient measurement of sodium currents in small cells (Hamill *et al.*, 1981). These procedures have been adapted for studies of neuroblastoma cells by Huang *et al.* (1982). In our experiments, patch pipettes with a tip resistance of 0.4 to 1.0 MΩ are pressed against the cell membrane and a seal of 0.1–5 GΩ is formed under suction. The solution within the pipette contains 160 m*M* CsF. In this solution, a conductive pathway forms under maintained suction, making the inside of the patch pipette electrically continuous with the cytoplasm. Ionic currents measured between the pipette and bath record ion movement through the whole surface membrane of the cell. Outward currents through K^+ channels are blocked by the Cs^+ in the patch pipette. Therefore, depolarization from a negative holding potential elicits a current that is completely blocked by tetrodotoxin and therefore represents Na^+ movement into the cells through sodium channels. Typically, cells were held at a membrane potential of −75 mV, hyperpolarized to −105 mV for 90 msec, and then depolarized to various membrane potentials for 10 msec to activate sodium channels. Figure 4A illustrates sodium currents in response to depolarizations to potentials at 15-

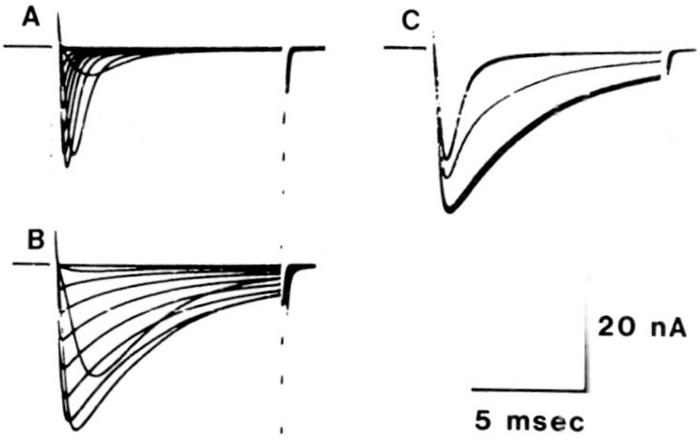

FIG. 4. Sodium currents in N18 cells in the presence and absence of scorpion toxin. A whole cell patch was established as described by Huang *et al.* (1982) with 160 m*M* CsF within the pipette. The cell was held at −75 mV. For each record, the cell was hyperpolarized to −105 mV for 90 msec to reverse inactivation of the channel and then depolarized to the pulse potential for 10 msec. (A) Sodium currents for step polarizations to −60 mV to +90 mV in 15-mV intervals measured 60 min after making patch. (B) Sodium currents evoked by the same pulses in the presence of 100 n*M* scorpion toxin. (C) Time course of scorpion toxin effect was determined by measuring sodium currents evoked by depolarization to 0 mV before addition of the toxin or 3.0 and 8.0 min following addition.

mV intervals from −60 to +90 mV. The sodium currents are well resolved throughout their time course by the techniques used.

For each of the currents in experiments like those in Fig. 4A, the equivalent sodium conductance (G_{Na}) was calculated and plotted vs membrane potential for N18 cells and the *scorpion toxin-resistant* clones LV10 and LV30. These curves were used to calculate the maximum sodium conductance (G_{Na}) and the membrane potential giving half-maximal G_{Na} ($E_{G=0.5}$). These values are given in Table V. Maximum G_{Na}, and therefore the number of sodium channels, in the mutant cells is reduced. However, the values of $E_{G=0.5}$ are not altered significantly (Table V) and the G_{Na} vs V curves are nearly superimposable over the full range of membrane potentials studied (data not shown). Similarly, no difference in the time course of activation was noted (Table V). These results show that the voltage dependence of sodium channel activation is not altered in *scorpion toxin-resistant* mutants.

The decline in sodium current during a maintained depolarization reflects the process of inactivation. The voltage dependence of steady-state inactivation was measured by changing the membrane potential from a holding potential of −75 mV to various potentials for 90 msec and then activating all channels that are not inactivated by a depolarization to +7.5 mV (Hodgkin and Huxley, 1952). During the first voltage step, some sodium channels activate and then inactivate while others inactivate directly without passing through the open state. The extent of inactivation depends upon the membrane potential. The fraction remaining in the resting state is then determined by recording sodium current

TABLE V
ELECTROPHYSIOLOGICAL PROPERTIES OF SODIUM CHANNELS IN NORMAL AND NEUROTOXIN-RESISTANT NEUROBLASTOMA CELLS

Parameter	Cell lines		
	N18	N10	LV30
Maximum conductance, G_{Na} (mS/cm^2)	40 ± 31	28 ± 12	24 ± 9
Half activation potential, $E_{G=0.5}$ (mV)	−9 ± 4	−11 ± 4	−7 ± 2
Half inactivation potential, $E_{h=0.5}$ (mV)	−66 ± 5	−69 ± 6	−62 ± 6
Time to peak current at 15 mV, (msec)	0.54 ± 0.06	0.50 ± 0.04	0.59 ± 0.12

during a further depolarization. The voltage dependence of steady-state inactivation was similar for N18, LV10, and LV30 over the whole voltage range tested. The potential giving half-maximum inactivation ($E_{h=0.5}$) was −62 mV ± 5 mV for N18, −76 mV ±6 mV for LV10, and −62 mV ± 5 mV for LV30 (Table V). Thus, sodium channels in *scorpion toxin-resistant* neuroblastoma cells inactivate with a voltage dependence that is similar to that of N18 cells.

The effect of *Leiurus quinquestriatus* scorpion toxin on sodium currents in N18 cells and *scorpion toxin-resistant* neuroblastoma cells was studied to verify the conclusion from biochemical studies that these cells have reduced affinity for scorpion toxin. Figure 4B and C illustrate the effect of scorpion toxin on the time course of the sodium current in N18 cells. The increase in sodium current is little affected, but the currents are greatly prolonged in the presence of the toxin. The decay of the sodium current during a maintained depolarization is approximately exponential in N18 cells as in other preparations. Table VI presents the time constants for decay of sodium currents in N18, LV10, and LV30. In each case, a saturating concentration of scorpion toxin increases τ from approximately 0.5 to 4–5 msec. The concentration dependence of scorpion toxin action was measured by determining the fraction of sodium current remaining 3 msec after the peak. For unmodified channels, less than 2% remains; for scorpion toxin-modified channels, approximately 70% remains. Concentration–effect curves for *Leiurus*

TABLE VI
EFFECT OF SCORPION TOXIN ON SODIUM CURRENT IN NORMAL AND NEUROTOXIN-RESISTANT NEUROBLASTOMA CELLS

Parameter	Cell lines		
	N18	LV10	LV30
I_{Na} decay time (msec)			
Control	0.55 ± 0.13	0.45 ± 0.08	0.58 ± 0.12
+200 n*M* scorpion toxin	5.7 ± 1.1	4.9 ± 0.7	3.7 ± 0.6
K_D for scorpion toxin (n*M*)	1.7	24	5.4

toxin action show that higher concentrations of toxin are required to slow inactivation of sodium channels in LV10 and LV30 (data not shown). Apparent K_D values increase from 1.7 n*M* for N18 cells to 24 n*M* and 5.4 n*M* for LV10 and LV30, respectively (Table VI). These results confirm the conclusion from biochemical studies that *scorpion toxin-resistant* cell lines have sodium channels with reduced affinity for scorpion toxin. In addition, our results show that this change in sodium channel properties is specific and does not markedly affect the normal function of the channel.

E. Biochemical Analysis of Neurotoxin-Resistant Cell Lines

The emerging information on the molecular properties of sodium channels has allowed us to begin to examine the molecular basis of the different neurotoxin-resistant phenotypes. Our initial approach was to develop methods to identify the polypeptide corresponding to the α subunit in normal and neurotoxin-resistant neuroblastoma clones (Costa and Catterall, 1982). Neuroblastoma cells were grown in the presence of [^{35}S]methionine to label the cellular protein pool to high specific activity. Cells were then solubilized in Triton X-100 and a membrane glycoprotein fraction was prepared by chromatography on wheat germ agglutinin immobilized on Sepharose beads. This purified glycoprotein fraction was then analyzed by SDS gel electrophoresis and by two-dimensional isoelectric focusing/SDS gel electrophoresis. A glycoprotein band of 270,000 Da was identified as the α subunit of the sodium channel in N18 cells by comparison with standards. This band was missing in *sodium channel-deficient* clones such as clone LV9. In two-dimensional gels, a protein spot with M_r = 270,000 and p*I* = 5.8 was identified as the sodium channel from N18 cells. This spot was missing from *sodium channel-deficient* clones and was present in reduced amount in *scorpion toxin-resistant* clones (Costa and Catterall, 1982). The spot corresponding to the α subunit was broad in the isoelectric focusing dimension suggesting microheterogeneity in this protein. No change in isoelectric point was observed in *scorpion toxin-resistant* clones within the resolution of these methods. These experiments indicate that the sodium channel-deficient phenotype arises from failure to successfully synthesize and incorporate the α subunit, and possibly also the other subunits, of the sodium channel into the cell membrane. The site of the lesion in the *scorpion toxin-resistant* cells remains uncertain.

V. CONCLUSION

Voltage-sensitive ion channels are unique in their functional response to small changes in membrane voltage. It is likely that these membrane transport proteins contain special structural features that underlie their marked voltage sensitivity. The voltage-sensitive sodium channel is the first of these macromolecules to be examined at the molecular level. The protein has been purified to homogeneity from rat brain and shown to consist of a stoichiometric complex of 316,000 Da containing three nonidentical subunits: α, β1, and β2. This purified protein is sufficient to mediate the functional activities of the sodium channel when incorporated into phospholipid vesicles of appropriate composition. Somatic cell genetic analysis of the structural and functional properties of the sodium channel has begun with the selection and characterization of variant neuroblastoma clones with missing or altered sodium channels. Variants with sharply reduced levels of sodium channels and with sodium channels having reduced affinity for scorpion toxin have been identified. Localization of the molecular defects responsible for these phenotypes may identify important functional domains of the sodium channel. It is hoped that the combination of these biochemical and somatic cell genetic approaches with the new methods of molecular genetics will lead to a detailed understanding of the molecular basis of electrical excitability.

REFERENCES

Albuquerque, E. X., and Daly, J. W. (1976). Batrachotoxin, a selective probe for channels modulating sodium conductances in electrogenic membranes. *In* "Receptors and Recognition" (P. Cuatrecasas, ed.), pp. 299–338. Chapman & Hall, London.

Baker, R. M., Brunette, D. M., Mankovitz, R., Thompson, L. H., Whitmore, G. F., Siminovitch, L., and Till, J. E. (1974). Ouabain-resistant mutants of mouse and hamster cells in culture. *Cell* **1,** 9–21.

Barhanin, J., Giglio, J. R., Leopold, P., Schmid, A., Sampaio, S. V., and Lazdunski, M. (1982a). *Tityus serrulatus* venom contains two classes of toxins. *J. Biol. Chem.* **257,** 12553–12558.

Barhanin, J., Schmid, A., Lombet, A., Wheeler, K. P., and Lazdunski, M. (1982b). Molecular size of different neurotoxin receptors on the voltage-sensitive Na^+ channel. *J. Biol. Chem.* **258,** 700–702.

Beneski, D. A., and Catterall, W. A. (1980). Covalent labeling of protein components of the sodium channel with a photoactivable derivative of scorpion toxin. *Proc. Natl. Acad. Sci. U.S.A.* **77,** 639–643.

Benzer, T. I., and Raftery, M. A. (1973). Solubilization and partial characterization of the tetrodotoxin binding component from nerve axons. *Biochem. Biophys. Res. Commun.* **51,** 939–944.

Bourne, H. R., Coffino, P., and Tomkins, G. M. (1975). Selection of a variant lymphoma cell deficient in adenylate cyclase. *Science* **187,** 750–752.

Brown, G. B., Tieszen, S. C., Daly, J. W., Warnick, J. E., and Albuquerque, E. X. (1981). Batrachotoxinin-A 20-α-Benzoate: A new radioactive ligand for voltage sensitive sodium channels. *Cell. Mol. Neurobiol.* **1,** 19–40.

Cahalan, M. D. (1975). Modification of sodium channel gating in frog myelinated nerve fibres by *Centruroides sculpturatus* scorpion venom. *J. Physiol.* (*London*) **244,** 511–534.

Catterall, W. A. (1977). Activation of the action potential Na^+ ionophore by neurotoxins. An allosteric model. *J. Biol. Chem.* **252,** 8669–8676.

Catterall, W. A. (1979). Binding of scorpion toxin to receptor sites associated with sodium channels in frog muscle. *J. Gen. Physiol.* **74,** 375–391.

Catterall, W. A. (1980). Neurotoxins that act on voltage-sensitive sodium channels in excitable membranes. *Annu. Rev. Pharmacol. Toxicol.* **20,** 15–43.

Catterall, W. A., Morrow, C. S., and Hartshorne, R. P. (1979). Neurotoxin binding to receptor sites associated with voltage-sensitive sodium channels in intact, lysed, and detergent-solubilized brain membranes. *J. Biol. Chem.* **254,** 11379–11387.

Catterall, W. A., Morrow, C. S., Daly, J. W., and Brown, G. B. (1981). Binding of batrachotoxinin A 20-α-benzoate to a receptor site associated with sodium channels in synaptic nerve ending particles. *J. Biol. Chem.* **256,** 8922–9927.

Conti, F., Hille, B., Neumcke, B., Nonner, W., and Stampfli, R. (1976). Conductance of the sodium channel in myelinated nerve fibres with modified sodium inactivation *J. Physiol. (London)* **262,** 729–742.

Costa, M. R., and Catterall, W. A. (1982). Characterization of variant neuroblastoma clones with missing or altered sodium channels. *Mol. Pharmacol.* **22,** 196–203.

Couraud, F., Jover, E., Dubois, J. M., and Rochat, H. (1982). Two types of scorpion toxin receptor sites, one related to the activation and the other to the inactivation of the action potential sodium channel. *Toxicon,* **20,** 9–16.

Fenwick, R. G., and Caskey, C. T. (1975). Mutant chinese hamster cells with a thermosensitive hypoxanthine-guanine phosphoribosyl transferase. *Cell* **5,** 115–122.

Frelin, C., Vigne, P., and Lazdunski, M. (1981). The specificity of the sodium channel for monovalent cations. *Eur. J. Biochem.* **119,** 437–442.

Goldin, S. M., Rhoden, V., and Hess, E. J. (1980). Molecular characterization, reconstitution, and "transport-specific fractionation" of the saxitoxin binding protein/Na^+ gate of mammalian brain. *Proc. Natl. Acad. Sci. U.S.A.* **77,** 6884–6888.

Hamill, O. P., Marty, A., Neher, E., Sakmann, B., and Sigworth, F. J. (1981). Improved patch-clamp techniques for high resolution current recording from cells and cell-free patches. *Pflugers Arch. Gesamte Physiol.* **391,** 85–100.

Hartshorne, R. P., and Catterall, W. A. (1981). Purification of the saxitoxin receptor of the sodium channel from rat brain. *Proc. Natl. Acad. Sci. U.S.A.* **78,** 4620–4624.

Hartshorne, R. P., and Catterall, W. A. (1984). The sodium channel from rat brain. Purification and subunit composition. *J. Biol. Chem.* **259,** 1667–1675.

Hartshorne, R. P., Coppersmith, J., and Catterall, W. A. (1980). Size characteristics of the solubilized saxitoxin receptor of the voltage-sensitive sodium channel from rat brain. *J. Biol. Chem.* **255,** 10572–10575.

Hartshorne, R. P., Messner, D. J., Coppersmith, J. C., and Catterall, W. A. (1982). The saxitoxin receptor of the sodium channel from rat brain. *J. Biol. Chem.* **257,** 13888–13891.

Henderson, R., and Wang, J. H. (1972). Solubilization of specific tetrodotoxin-binding component from garfish olfactory nerve membrane. *Biochemistry* **11,** 4565–4569.

Hille, B. (1971). The permeability of the sodium channel to organic cations in myelinated nerve. *J. Gen. Physiol.* **58,** 599–619.

Hille, B. (1972). The permeability of the sodium channel to metal cations in myelinated nerve. *J. Gen. Physiol.* **59,** 637–658.

Hodgkin, A. L., and Huxley, A. F. (1952). A quantitative description of membrane current and its application to conduction and excitation in nerve. *J. Physiol. (London)* **117,** 500–544.

Huang, L. M., Catterall, W. A., and Ehrenstein, G. (1979). Comparison of ionic selectivity of batrachotoxin-activated channels with different tetrodotoxin dissociation constants. *J. Gen. Physiol.* **73,** 839–854.

Huang, L. M., Moran, N., and Ehrenstein, G. (1982). Batrachotoxin modifies the gating kinetics of sodium channels in internally perfused neuroblastoma cells. *Proc. Natl. Acad. Sci. U.S.A.* **79,** 2082–2085.

Jover, E., Martin-Moutot, N., Couraud, F., and Rochat, H. (1978). Scorpion toxin: specific binding to synaptosomes. *Biochem. Biophys. Res. Commun.* **85,** 377–382.

Jover, E., Couraud, F., and Rochat, H. (1980). Two types of scorpion neurotoxins characterized by their binding to two separate receptor sites on rat brain synaptosomes. *Biochem. Biophys. Res. Commun.* **95,** 1607–1614.

Khodorov, B. I. (1978). Chemicals as tools to study nerve fiber sodium channels; effects of batrachotoxin and some local anesthetics. *Membr. Transp. Processes* **2,** 153–174.

Krueger, B. K., and Blaustein, M. P. (1980). Sodium channels in presynaptic nerve terminals. Regulation by neurotoxins. *J. Gen. Physiol.* **76,** 287–313.

Krueger, B. K., Ratzlaff, R. W., Strichartz, G. R., and Blaustein, M. P. (1979). Saxitoxin binding to synaptosomes, membranes, and solubilized binding sites from rat brain. *J. Membr. Biol.* **50,** 287–310.

Levinson, S. R., and Ellory, J. C. (1973). Molecular size of the tetrodotoxin binding site estimated by irradiation inactivation. *Nature (London) New Biol.* **245,** 122–123.

Malysheva, M. K., Lishko, U. K., and Chagovetz, A. M. (1980). The association of the tetrodotoxin-sensitive sodium-selective ionophore of brain membranes with liposomes. *Biochim. Biophys. Acta* **602,** 70–76.

Meves, H., Rubly, N., and Watt, D. D. (1982). Effect of toxins isolated from the venom of the scorpion *Centruroides sculpturatus* on the Na currents of the node of Ranvier. *Pflugers Arch.* **393,** 56–62.

Narahashi, T. (1974). Chemicals as tools in the study of excitable membranes. *Physiol. Rev.* **54,** 813–889.

Ray, R., Morrow, C. S., and Catterall, W. A. (1978). Binding of scorpion toxin to receptor sites associated with voltage-sensitive sodium channels in synaptic nerve ending particles. *J. Biol. Chem.* **253,** 7307–7313.

Ritchie, J. M., and Rogart, R. B. (1977). The binding of saxitoxin and tetrodotoxin to excitable tissue. *Rev. Physiol. Biochem. Pharmacol.* **29,** 1–50.

Ross, E. M., Howlett, A. C., Ferguson, K. M., and Gilman, A. G. (1978). Reconstitution of hormone-sensitive adenylate cyclase activity with resolved components of the enzyme. *J. Biol. Chem.* **253,** 6401–6411.

Sigworth, F. J. (1980). The conductance of sodium channels under conditions of reduced current at the node of Ranvier. *J. Physiol. (London)* **307,** 131–142.

Sigworth, F. J., and Neher, E. (1980). Single Na^+ channel currents observed in cultured rat muscle cells. *Nature (London)* **287,** 447–449.

Talvenheimo, J. A., Tamkun, M. M., and Catterall, W. A. (1982). Reconstitution of neurotoxin-stimulated sodium transport by the voltage-sensitive sodium channel purified from rat brain. *J. Biol. Chem.* **257,** 11868–11871.

Tamkun, M. M., and Catterall, W. A. (1981). Ion flux studies of voltage-sensitive sodium channels in synaptic nerve ending particles. *Mol. Pharmacol.* **19,** 78–86.

Tamkun, M. M., Talvenheimo, J. A., and Catterall, W. A. (1984). The sodium channel from rat brain: Reconstitution of neurotoxin-activated ion flux and scorpion toxin binding from purified components. *J. Biol. Chem.,* **259,** 1676–1688.

Villegas, R., and Villegas, G. M. (1981). Nerve sodium channels incorporation in vesicles. *Annu. Rev. Biophys. Bioeng.* **10,** 387–419.

Wang, G. K., and Strichartz, G. (1982). Simultaneous modifications of sodium channel gating by two scorpion toxins. *Biophys. J.* **40,** 175–179.

Weigele, J. B., and Barchi, R. L. (1982). Functional reconstitution of the purified sodium channel protein from rat sarcolemma. *Proc. Natl. Acad. Sci. U.S.A.* **79,** 3651–3655.

West, G. J., and Catterall, W. A. (1979). Selection of variant neuroblastoma clones with missing or altered sodium channels. *Proc. Natl. Acad. Sci. U.S.A.* **76,** 4136–4140.

Part III

Transport Systems

Chapter 6

The Histidine Transport System of *Salmonella typhimurium*

GIOVANNA FERRO-LUZZI AMES

Department of Biochemistry
University of California
Berkeley, California

Cells of *Escherichia coli* and *Salmonella typhimurium* can transport substrates either by active transport systems, by facilitated diffusion, or by group translocation. Active transport, which allows accumulation of substrate in unmodified form against a concentration gradient, can occur through two main classes of transport systems. The first class is typically represented by the β-galactoside permease (i.e., the M protein, coded for by the *lacY* gene of the lactose operon), which is a relatively simple system involving only a single membrane-bound protein; such systems are energized by an ion gradient. The second class includes more complex systems, which are sensitive to mild osmotic shock, and require three or more protein components to operate. Transport systems belonging to the second class are referred to either as ''periplasmic permeases'' or as ''shock-sensitive permeases'' because one of their constituents is located in the periplasm and can be released by osmotic shock (Heppel, 1971).

ISBN 0-12-153323-9

I. THE PERIPLASM

The existence of a separate compartment in these gram-negative cells, the periplasm, was proposed by Heppel and collaborators (Heppel, 1971), and the term has been used to describe an operationally defined space, i.e., that area of the cell which contains proteins that can be released by a mild, cold osmotic shock procedure. This consists essentially of a plasmolysis step with sucrose in the presence of EDTA, followed by resuspension in cold distilled water. Approximately 15% of the total cell protein is released into the medium by osmotic shock, leaving the cells viable, though unable to perform the functions that depend upon the lost periplasmic proteins. These proteins are usually involved in functions that are cell surface related, such as transport or degradation of metabolizable compounds which are too large or too highly charged to enter the cell intact (e.g., RNase and a variety of phosphatases are among the periplasmic proteins). The periplasm is usually thought to be physically located between the inner and outer membranes.

The first indication of the involvement of a soluble substrate-binding protein in transport was obtained by Pardee who discovered the periplasmic sulfate-binding protein while studying sulfate transport (Pardee and Prestidge, 1966). Since then numerous substrates have been shown to utilize periplasmic systems in gram-negative organisms.

II. MULTIPLICITY OF TRANSPORT SYSTEMS

It is an interesting and, at the same time, frustrating problem that many substrates, possibly all, make use of more than one transport system in entering the cell. It is, therefore, imperative to resort to genetic tools to sort out the variety of overlapping permeases functioning on any one solute (Ames, 1964, 1972).

At least five separate transport systems are involved in the uptake of histidine by *Salmonella typhimurium* (Ames, 1972). One of these, the high-affinity histidine permease, is a shock-sensitive system. Another permease makes use of a different periplasmic protein (LAO, see below) and of some of the components of the high-affinity histidine permeases. Little is known about the other systems. This pattern of multiplicity, which was initially established for histidine and aromatic amino acids in *S. typhimurium* (Ames, 1964; Ames and Roth, 1968), has been subsequently shown to be a general phenomenon and also to apply to substrates other than amino acids (see reviews in Rosen, 1978; Payne, 1980). Isolation and characterization of transport mutants would allow the distinction to be made between the numbers and kinds of permeases; this can be achieved using a variety of techniques. Those that have been used in the study of histidine transport (a system that has received a very extensive genetic characterization) are generally applicable to the analysis of any other transport system. Thus, mutants can be selected by (1) their resistance to inhibitory analogs transported by the system (Ames, 1964; Ames and Roth, 1968), (2) their improved ability to utilize an essential metabolite which is a poor substrate of the system (Krajewska-Grynkiewicz *et al.*, 1971), or (3) loss of the ability to use an essential metabolite that is transported by the system (Ames and Lever, 1970). In addition, transposons (e.g., Tn10) can be used to create insertions and deletions in transport genes (Noel and Ames, 1978).

III. THE HIGH-AFFINITY HISTIDINE PERMEASE

The high-affinity histidine permease of *S. typhimurium* is the best characterized system of several distinct histidine permeases that function in this organism. An analogous system has been identified in *E. coli* (Ardeshir and Ames,

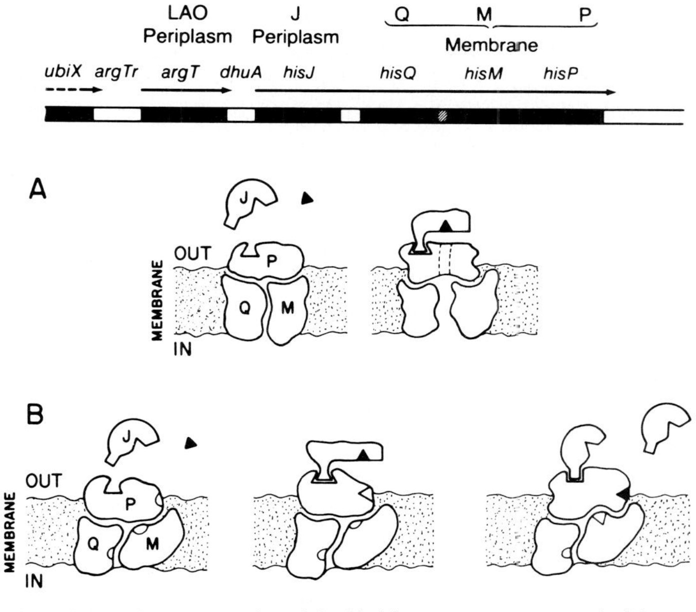

FIG. 1. (Top) Schematic representation of the histidine transport operon and of the gene (*argT*) coding for the LAO protein. The sites called *argTr* and *dhuA* are the regulatory regions for *argT* and for the histidine transport operon, respectively. The areas in black indicate the structural genes. The areas in white indicate the regulatory regions or regions at the end of operons. The horizontal arrows indicate direction of transcription and presumed size of messenger RNAs (though these have not been characterized yet). (Bottom) Models for transport. See text for discussion of details. The solid triangle represents histidine. J is represented as interacting with P upon having bound histidine. In (B) potential histidine-binding sites on Q, M, and P are represented by hemispherical indentations which are sequentially "activated" (to triangular shapes). Only the initial transfer of histidine from J to P with concomitant conformational changes is illustrated. Transfer of histidine to M and on to Q can be imagined to occur by a similar series of events.

1980). A detailed genetic map of the histidine transport operon has been constructed (Ames and Spudich, 1976). The operon, located at 48.5 minutes on the *Salmonella* linkage map, consists of four genes, *hisJ, hisQ, hisM,* and *hisP,* which are under the control of a promoter/operator locus, *dhuA*. Figure 1 (top) shows a schematic representation of the genetic organization of this permease. Expression of this transport operon, and that of several other transport systems, is regulated by nitrogen availability. The *hisJ* gene codes for a periplasmic histidine-binding protein (molecular weight 25,000), which has been sequenced (Hogg, 1981). The hisP, hisQ, and hisM gene products are membrane-bound proteins (Ames and Nikaido, 1978; Higgins *et al.*, 1982b). In recent years, the advent of recombinant DNA technology has allowed us to obtain the nucleotide sequence of the entire histidine transport operon (Higgins *et al.*, 1982b). All of the known mutations in this transport system map in this operon. Though initially surprising, this multiplicity of components—two or more membrane bound in addition to the periplasmic one—has been found to be true of periplasmic permeases for very different types of substrates: e.g., the branched-chain amino acids (Oxender *et al.*, 1980), maltose (Shuman, 1982a), galactose (Robbins *et al.*, 1976), and arabinose (Clark and Hogg, 1981).

IV. BIOCHEMICAL CHARACTERIZATION OF TRANSPORT COMPONENTS

The periplasmic histidine-binding protein has been characterized thoroughly. Among its most important properties is its ability to undergo a definite change in conformation upon binding its substrate (Robertson *et al.*, 1977; Zukin *et al.*, 1984). All other periplasmic transport proteins that have been examined for this property also undergo a conformational change upon binding of their respective substrates: the galactose-, arabinose-, ribose-, maltose, glutamine-, and leucine–isoleucine–valine-binding proteins (Furlong and Schellenberg, 1980). Thus, a substrate-induced conformational change seems to be an essential aspect of the mechanism of action of these proteins.

In an effort to understand which part of the J molecule might be involved in what aspect of transport, we were able to show that the histidine-binding protein contains two separate domains, one of which is responsible for binding the transported substrate, histidine, while the other appears to be involved in an interaction with at least one of the membrane-bound components, the P protein. This was accomplished by searching for and identifying a mutant in *hisJ* which produces an altered J protein unable to function in transport, but still able to bind histidine with unaltered affinity. Such a mutation, therefore, has altered a domain of the J molecule which, though essential for transport function, is not involved in the formation of the histidine-binding site (Kustu and Ames, 1974).

We postulated that the site where the mutation has occurred is involved in interacting with the P protein because a pseudorevertant correcting the transport defect could be isolated which maps in the *hisP* gene (Ames and Spudich, 1976).

The membrane-bound components of the histidine permease are present in much smaller amounts than the soluble one, and they are thought to interact with each other, thus forming a membrane-bound complex. This complex is unable to function in transport in the absence of the binding protein. Evidence for the existence of such a complex comes from genetic complementation studies which indicated no, or aberrant, complementation between mutants in the genes coding for the membrane-bound components (Higgins *et al.*, 1982b). No solid biochemical evidence proving the existence of such a membrane-bound complex is yet available for any periplasmic system.

Since the membrane-bound components are present in very small amounts and

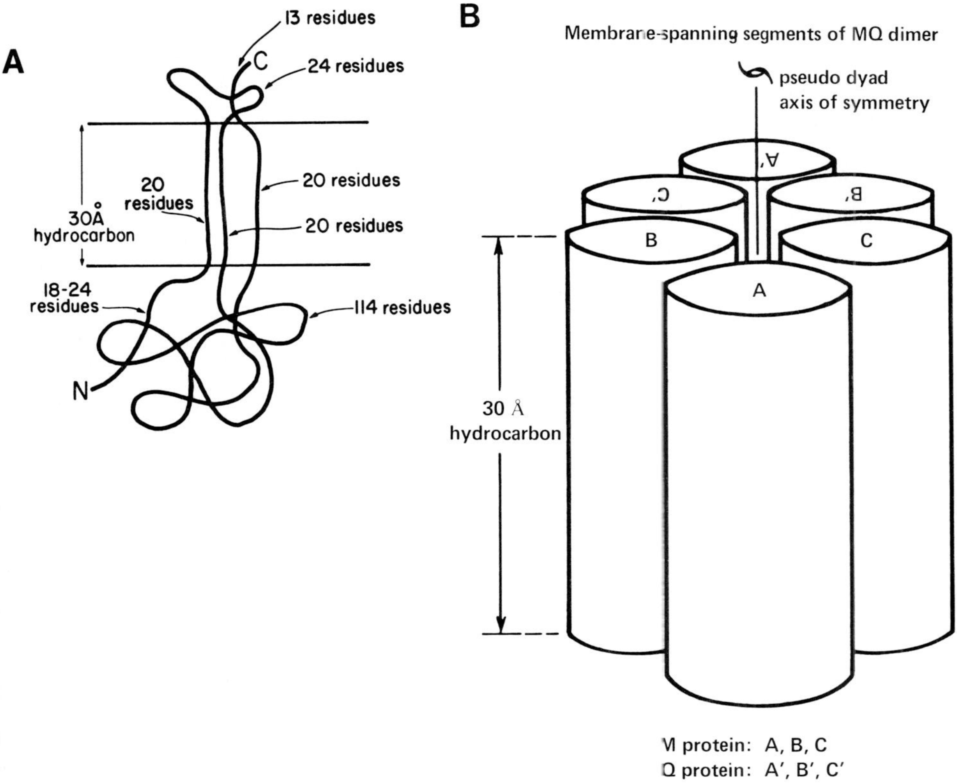

FIG. 2. Hypothetical distribution of Q or M across the membrane. (A) Three "membrane spanners" were identified by SOAP analysis (Kyte and Doolittle, 1982) in an identical location in either Q or M. (B) The three spanners of Q and M are represented as forming a pseudodimer in the membrane.

no assay for the activity in a cell extract is available, they have resisted all attempts at purification up to now. What little is known of them is derived from inspection of their sequence. The Q and M proteins are very hydrophobic and yield upon analysis of their hydropathicity by the SOAP program (Kyte and Doolittle, 1982) identical patterns of membrane-crossing sequence structures (this analysis was kindly performed by Jack Kyte). Each protein contains three "membrane-spanners" located in correspondingly identical locations (Fig. 2A). The two proteins are thus structurally similar enough that they may come together to form a psuedodimer in the membrane (Fig. 2B). Their derived amino acid sequences show a significant homology to each other, and they probably arose by duplication from an ancestral gene (Fig. 3). The P protein, on the other hand, is very hydrophilic, in fact, its level of hydrophilicity is indistinguishable from that of soluble, globular proteins (Gilson *et al.*, 1982). None of the genes coding for each of these three proteins contains a signal sequence. This suggests that while the hydrophobic Q and M proteins are embedded in the membrane as integral membrane proteins, P is a peripheral protein possibly located on the interior of the cytoplasmic membrane. If the latter were true, it would represent a serious problem for the present design of the models for the mechanism of action, which locate the P protein in direct contact with the J protein (and therefore on the outer surface of the membrane) on the basis of genetic data (Ames and Spudich, 1976).

```
hisM  M I E I I Q E Y W K S L L W T D G Y R F T G V A I T L W L L
hisQ                      M L Y G F S G V I L Q G A I V T L E L A

      I S S V V M G G L L A V I L A V G R V S S N K F I R F P I W
      L S S V V L A V L I G L V G A G A K L S Q N R V T G L I F E

      L F T Y I F R G T P L Y V Q   L L V F Y S G M Y T L G I V L
      G Y T T L I R G V P D L V L M L L I F Y G L Q I A L N V V T

      G T D L L N A F F R S G L N C T V L A L T L N T C A Y T T E
      D S L G I D Q I D I D P M V A G I I T L G F I Y G A T F T E

      I F A G A I R S V P H G E I E A A R A Y G F S S F K M Y R C
      T F R G A F M A V P K G H I E A A T A F G F T H G Q T F R R

      I I L P S A L R I A L P A Y S N G V I L M L H S T A L A F T
      I M F P A M M R Y A L P G I G N N W Q V I L K A T A L V S L

      A T V P D L L K I A R D I N S A T Y Q P F T A F G I A A V L
      L G L E D V V K A T Q L A G K S T W E P F Y F A V V C G L I

      Y L L I S Y V L I S L F R R A E R R W L Q H V S S K
      Y L V F T T V S N G V L L L L E R R Y S V G V K R A D L
```

FIG. 3. Homology between *hisQ* and *hisM*. The amino acid sequence has been derived from the nucleic acid sequence (Higgins *et al.*, 1982b). This computer-assisted comparison was performed by T. Farrah, with the help of Dr. R. Doolittle and using his programs. Boxed residues are identical in the two sequences.

V. POSSIBLE MECHANISMS OF ACTION

One of the goals of our research is to reconstitute the purified components into vesicles or liposomes, which would allow us to investigate closely the molecular mechanism of action of this transport system. Though this may now become feasible with the development of a method for the reconstitution of periplasmic systems in bacterial membrane vesicles (Hunt and Hong, 1981), it certainly is not yet a tool to be used routinely. We have attempted anyway to design transport models using the available information and these are shown in Fig. 1A and B. In the "pore" model (Fig. 1A), the binding of histidine to J causes a conformational change that allows interaction of the J-histidine complex with P: An additional conformational change within the P, Q, M complex allows a pore to be formed within or between the proteins of the complex (or both), allowing passage of histidine to the cell interior. Alternatively, the "binding-site" model shown in Fig. 1B postulates a more active role for the membrane-bound proteins, involving a histidine-binding site on each of the membrane-bound components (although, of course, not all of the membrane components need have such a site). While the initial step would be the same, upon interaction of the J–histidine complex with the membrane proteins, these would respond by sequentially "activating" one or more histidine-binding sites which would transport histidine in a bucket brigade-like manner through the membrane. In either case we must postulate that the interaction of the J–histidine complex with the membrane-bound proteins should be such that histidine is very close to the pore or to the next histidine-binding site, otherwise it would just diffuse into the periplasm as soon as it was released from the periplasmic component. In other words, the most probable function of the binding protein is that of "trapping" histidine in a bound form and delivering it directly to the next site, without allowing it to return to a freely diffusible state. The numerical excess of the periplasmic component as compared to the membrane components (a fact that has been observed in each system studied) then is understandable in the above terms as allowing an increase in the level of immobilization of the substrate in the periplasm. It is also possible to envisage models that combine features from each of the two models described.

It should also be noticed that, though not represented, an energy-coupling mechanism must be involved in this process of concentrative uptake. Information on the nature of the latter mechanism is still scanty: ATP, acetyl phosphate, or a compound derived from them have been postulated as energy donors for shock-sensitive transport systems (Hong *et al.*, 1979), though ion gradients have also been implicated (Plate, 1979). It should be mentioned that a posttranslational modification, such as phosphorylation or methylation of one of these transport components, has never been demonstrated.

It is too early to distinguish between these two models. However, it might be possible to do so by searching for mutants in the membrane proteins that have an *altered* spectrum of specificity for transport (Higgins *et al.*, 1982b). The wild-type histidine permease transports L-histidine efficiently; it also transports D-histidine and the histidine analog α-hydrazinoimidazolepropionic acid (HIPA) with lower affinity. The isolation of mutants in *hisQ, hisM,* or *hisP,* which *differentially* transport these substrates, suggests that the corresponding proteins have a substrate-binding site rather than being strictly part of a diffusion pore. We have found two types of such mutants: (1) *hisQ6699* transports L-histidine normally (K_m within 10% of normal) but does not transport D-histidine or HIPA at all within the limits of our measurements; (2) some *hisM* mutants (the "odd group", Ames *et al.*, 1977) result in the loss of transport of L- and D-histidine, and of HIPA, but also in an improved ability to transport L-histidinol. In biochemical terms such mutations would most easily (though not only) be interpreted as having caused an alteration in a substrate-binding site. Thus these mutants support the binding-site model. Further evidence for a substrate-binding site on the membrane components has been obtained for the maltose system, in which a mutant in one of the membrane-bound proteins has been characterized as allowing transport directly through the membrane, thus eliminating the need for the maltose-binding protein (Shuman, 1982b). The existence of a recognition site on the membrane has been postulated also by Rotman for the *mgl* transport system (Robbins *et al.*, 1976).

VI. CHANNELING FUNCTION OF MEMBRANE COMPONENTS

It is important to notice that an additional periplasmic component enters into the picture in this transport system. This is the lysine–arginine–ornithine-binding protein (LAO protein), which is immunologically related to the J protein (Kustu and Ames, 1974; Higgins and Ames, 1981). Like the J protein, the LAO protein also requires the Q, M, and P proteins in order to function in transport. However, when in combination with LAO, the system functions in the transport of arginine as a nitrogen source (Kustu and Ames, 1973). The LAO protein is believed to interact with the P protein in the course of arginine transport in an analogous manner to the J–P interaction during histidine transport (Fig. 4). The gene coding for the LAO protein, *argT,* is contiguous to the histidine transport operon (Fig. 1, top) and the sequences of the J and LAO proteins are highly homologous (70% identity, Higgins and Ames, 1981). The high homology, close location on the chromosome of these two genes, and the funneling of their function through the same membrane components strongly indicate that the two genes originated by tandem duplication of an ancestral gene, followed by divergent evolution. The funneling of the action of different periplasmic components

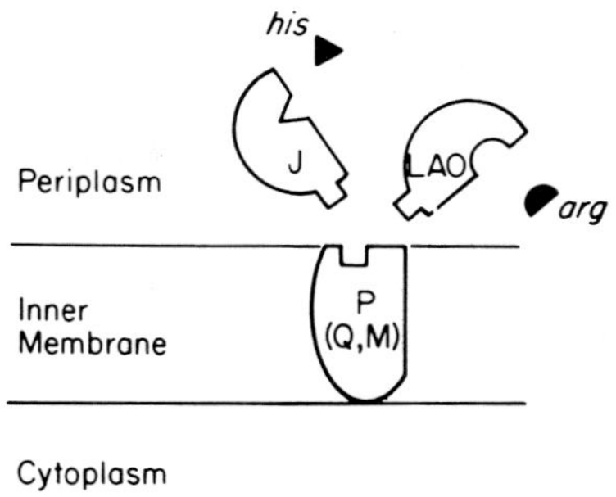

FIG. 4. The J and LAO proteins each interact with the membrane-bound transport components in a similar way. The two proteins are represented as having an identical domain which is the portion of each molecule responsible for the interaction with a site on the P protein in the membrane complex (P, M, Q).

through the same set of membrane-bound proteins is a property also of other periplasmic systems [e.g., the branched-chain amino acids system (Oxender *et al.*, 1980)], and it points to a commonalty of structure among these systems.

VII. REGULATION OF HISTIDINE AND ARGININE TRANSPORT

There are three good reasons for investigating the regulation of these systems. The first one is that little is known about the mechanisms regulating transport systems, with the exception of some systems transporting carbon sources (such as the *lac* system), which are regulated by the cyclic AMP–CRP mechanism for responding to carbon availability. The second is that genes within the histidine transport operon are expressed at widely different levels, thus suggesting that an intraoperon regulation is operative besides regulation at the promoter level. It would be useful to learn more of how operons regulate the level of individual genes. Finally, we know that both of these transport operons are regulated by nitrogen availability (Kustu *et al.*, 1979). Since this latter mechanism of regulation affects numerous cell functions (Magasanik, 1982), it would be useful to utilize the availability of our two characterized and completely sequenced nitrogen-regulated promoters for investigating this central regulatory phenomenon.

Since amino acid transport systems serve a similar function to the amino acid biosynthetic operons in supplying the cell with building blocks for protein synthesis, it seemed plausible that similar mechanisms might be involved in regulating expression of the transport and the biosynthetic operons. It is well known that several amino acid biosynthetic operons, including that for histidine, are regulated by a transcription attenuation mechanism (Yanofsky, 1981). However, neither *argTr* nor *dhuA* contains any of the features typical of the regulatory regions of the amino acid biosynthetic operons. (1) Neither region has a typical transcription termination sequence (apart from those responsible for termination of the preceding structural genes) appropriately located with respect to a series of overlapping dyad symmetries. (2) Although several sequences that could potentially encode leader peptides can be identified, none of these sequences seems likely to be of importance in regulation because (a) the starts of all of the potential peptides are located 5′ to the probable transcription start sites and could, therefore, not be transcribed or translated; (b) none of these peptides contain multiple tandem codons for amino acid(s) related to the operon. We conclude that attenuation, mediated by the premature termination of transcription, is not the mechanism by which expression of these genes is regulated. This is in agreement with the observation that neither the concentration of histidine in the medium, nor regulatory mutations (*hisR, hisS,* and *hisT*) known to increase the expression of the histidine biosynthetic operon (Brenner and Ames, 1971) has a major effect on the expression of the histidine transport genes.

Since the assay of transport components is difficult or impossible, in order to investigate further possible mechanisms of regulation we made use of Mu*dl* operon fusions which allowed us to measure transcriptional regulation by assaying for β-galactosidase (Casadaban and Cohen, 1979). Thus we confirmed that both the *dhuA* and the *argTr* promoters respond to nitrogen limitation, and the *argTr* promoter also responds to carbon limitation. It is interesting that these two forms of regulation are mutually exclusive, since growth under simultaneous limitation of carbon *and* nitrogen does not have an additive effect on levels of expression (Stern *et al.*, 1984a). No effect of either sulfur or phosphate starvation was detected. An examination of the nucleotide sequence of these two promoters revealed the following regions of interest. First, *argTr* contains a sequence with strong homology to the consensus sequence for the binding site for the CRP–cyclic AMP complex (O'Neill *et al.*, 1981). In agreement with the lack of carbon regulation at *dhuA*, the homologous location in *dhuA* contains a seqeunce with decreased homology with the consensus (Fig. 5). Interestingly,

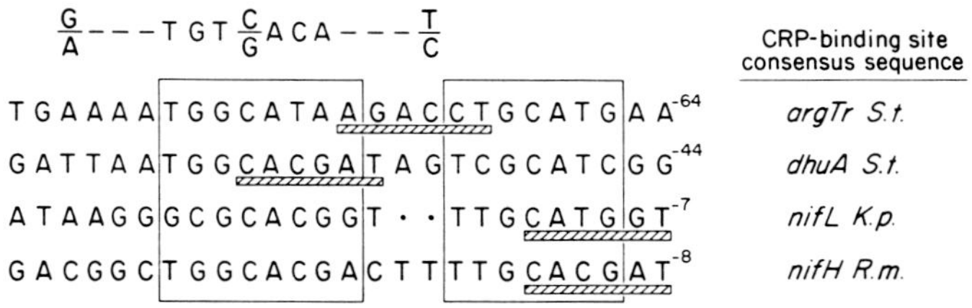

FIG. 5. Regulatory region of four nitrogen-regulated promoters. The boxed regions are homologous. The small hatched segments underneath the sequences indicate the locations of the Pribnow boxes. The consensus sequence for the CRP–cyclic AMP binding site at the top of the figure is aligned with the presumed carbon regulation site of *argTr*.

dhuA contains an additional region of good homology with the CRP protein-binding site located upstream from the −35 region of the promoter. However, we assume that this site is not functional since we detected no carbon regulation at *dhuA*.

Figure 5 shows also regions of strong homology to each other both in *dhuA* and *argTr* and in two other nitrogen-regulated promoters (Drummond *et al.*, 1983). This, then, is a possible site of action for proteins mediating nitrogen regulation. In agreement with the hypothesis is the fact that a promoter-up mutation in *dhuA*, *dhuA1*, eliminates nitrogen regulation at this promoter. This mutation has been sequenced and shown to occur at the Pribnow box, which is inside this region of homology. Also in agreement with the finding that the presumed nitrogen and carbon regulatory sites overlap is the fact that nitrogen regulation interfers with carbon regulation and vice versa, since these are not additive. The isolation and characterization of mutants affecting these forms of regulation will enable us to define further the regulatory sites.

One of the interesting aspects of operon regulation is at the level of intra-operonic gene expression. The *hisJ* gene is expressed at levels 10- to 30-fold higher than the downstream genes. A large intercistronic region separates *hisJ* from the downstream genes and it contains a large dyad symmetry (Higgins *et al.*, 1982a). Since it was possible that such structure might be responsible for the step-down level of expression of downstream genes, either by transcriptional or translational regulation, we examined the effect that its removal by deletion had on the expression of those genes (Stern *et al.*, 1984b). Neither transcription (as measured by using Mu*d1* fusions inserted downstream from the intercistronic region) nor translation (as measured by direct quantitation of the levels of the J and P proteins from two-dimensional gel electrophoretic resolution of the cell protein) was greatly affected: a 2-fold change was the maximum obtained by either measurement. Thus we feel that this intercistronic element is not involved in regulating step-down expression, which therefore may still be due to other unknown mechanism(s). It is interesting that the intercistronic element occurs repeatedly throughout the bacterial chromosome, in numerous but not identical copies bearing a high level of homology to each other (Stern *et al.*, 1984b). Since we feel that it is not involved in intraoperon regulation, it must have an important and different function. We have previously postulated its involvement in chromosomal rearrangement events such as duplication. It may also be involved in the bacterial nucleoid structure.

VIII. EVOLUTIONARY ASPECTS

A comparison of the characteristics of all the known periplasmic systems suggests that the underlying mechanism is possibly the same for all of them. All the permeases that have been studied extensively [i.e., the systems for histidine, the branched-chain amino acids (Landick *et al.*, 1980), maltose (Boos, 1982), and galactose (Robbins *et al.*, 1976)] have a similar composition, requiring one (or more, see below) periplasmic component and two or more membrane-bound components. The genes coding for the components are always closely linked on the chromosome, possibly forming an operon in all cases (though two divergent operons are known for the maltose transport system). Similar to the case with histidine, an interaction may occur between the periplasmic galactose-binding protein and a membrane-bound component (Strange and Koshland, 1976). It is interesting that the maltose-binding protein has been shown (Bavoil and Nikaido, 1981) to interact with an outer membrane protein (the LamB protein) which is a necessary component of this transport system. The requirement for an outer membrane protein is, possibly, a unique peculiarity of this system due to the necessity of handling high-molecular-weight substrates (maltodextrins) in addition to maltose (Shuman, 1982a).

An additional interesting complexity, which also seems to be shared by several periplasmic systems, is the duplication and divergent evolution of the gene coding for the periplasmic component. This has been shown clearly for the *hisJ* gene, which is highly homologous (70%) to the closely linked gene *argT*, coding for the LAO protein. A remarkably analogous situation exists for the branched-chain amino acid permease, where two periplasmic proteins are found: One is specific for leucine (LS protein); the other binds leucine, isoleucine, and valine (LIV protein). These proteins are highly homologous, require common membrane-bound proteins for their function, and their genes are closely linked on the chromosome (Landick *et al.*, 1980). It is very likely that these two genes also originated by duplication of a single ancestral gene. The incomplete characterization of most other systems does not allow a generalization to be formulated as far as duplication of genes for periplasmic components is concerned. However, scattered information supports the notion that duplication of the periplasmic component is a common characteristic of these transport systems: (a) the galactose- and the arabinose-binding proteins are antigenically related, they share a small but definite homology (Mahoney *et al.*, 1981), and their genes are located quite close to each other on the chromosome (Clark and Hogg, 1981); (b) two arginine-binding proteins, in addition to LAO (possibly the result of gene duplication?), are known to exist (Rosen, 1971); (c) a gene with unknown function but probably coding for a secreted protein might be part of the maltose transport operon (a duplication-originated binding protein for a maltose-related substrate?) (Clement and Hofnung, 1981).

Considering the complexity of their organization, it is possible that all the periplasmic systems have originated by duplication and divergence from a single ancestral system, perhaps already containing a duplication of the periplasmic component (Fig. 6). Each system would have evolved a different specificity

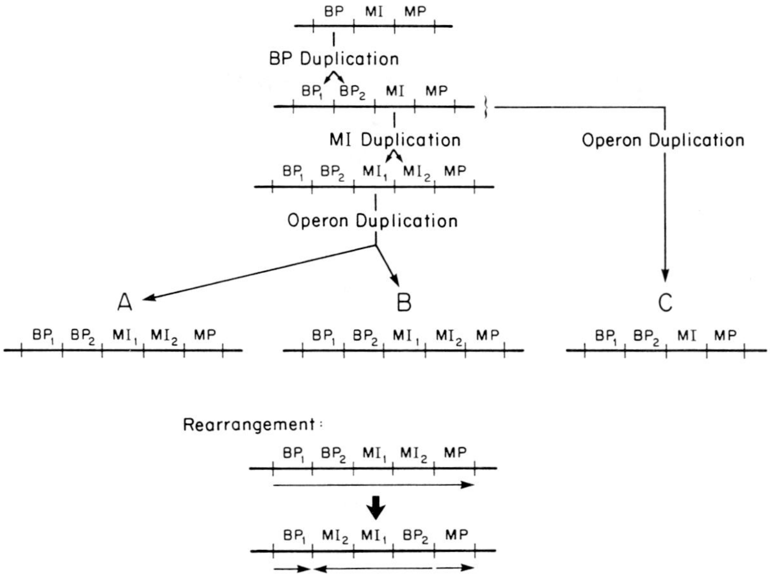

FIG. 6. Hypothetical evolutionary scheme for periplasmic transport systems. At the top is represented an ancestral operon coding only for three components: a soluble binding protein (BP), one integral membrane protein (MI), and a peripheral membrane protein (MP). Duplications, either of single genes within the operon, or of the entire operon, during various stages of evolution could create a variety of combinations of these genes. In addition, chromosomal rearrangements, such as the one depicted at the bottom of the figure, would have created additional individuality within the multitude of operons.

while retaining the same basic architecture. A search for homologies among parallel components of all these systems could answer this question. At the moment some tantalizing evidence is emerging. The sequence of one of the membrane-bound components of the histidine transport system, the P protein, has a clearly significant homology with the sequence of a membrane-bound component of the maltose transport system, the MalK protein (Gilson *et al.*, 1982). Comparison of all the available sequences of periplasmic components has shown marginally significant homologies between several completely unrelated binding proteins (Ames, Farrah, and Doolittle, unpublished observations). A more sophisticated statistical analysis will be necessary before the significance of such homologies can be determined. Also in favor of a structural and functional relationship between some of these proteins are the elegant studies of Quiocho and his collaborators on the X-ray structure of several periplasmic transport proteins, which showed in all cases strongly similar two-domain structures, even though the substrates transported by the proteins analyzed were quite different from each other (leucine–isoleucine–valine, arabinose, and sulfate) (Gilliland and Quiocho, 1981).

An obvious question arises if we are to conclude that all periplasmic systems arose from an ancestral system which already contained a duplication of the periplasmic component: Why is the homology between the two periplasmic components of one system greater than the homology between those same components and their duplicated versions as they appear in a system of completely different specificity? In other words, why would LAO resemble J more than either LAO or J resembles LS or LIV? A reasonable explanation is that LAO and J are constrained in their evolution by their need to interact with the same membrane component (the P protein), whereas LS and LIV presumably need to interact with their own specialized membrane component. This hypothesis is supported by the fact that within the J and LAO proteins there are segments that are much more highly conserved (>90% homology) than the overall protein sequence (70% homology). These conserved regions are believed to be involved in forming that domain of each protein molecule that is responsible for the interaction with the P protein (Fig. 2) (Higgins and Ames, 1981). Mutational alteration in one of these highly conserved regions of the J protein results in a loss of the ability of J to interact with P. The level of homology maintained between J and LAO is such that a chimeric protein resulting from a deletion fusing the amino-terminal half of LAO to the carboxy-terminal half of J functions perfectly normally with the membrane-bound components, though it maintains the substrate specificity of LAO (Higgins and Ames, 1981). Therefore, this system, and independently each of the other systems, would have evolved as a "package."

Thus, the evidence that a complex ancestral system would have spawned the present multiplicity of periplasmic permeases is tantalizing. However, more systems need to be fully characterized before serious formulation of models for their evolutionary history is possible.

IX. CONCLUSIONS

The study of several periplasmic systems acting upon different substrates has increasingly shown strong similarities among them. It is probable that our extensive analysis of the histidine transport system will continue to bear upon an understanding of periplasmic transport in general. Recombinant DNA technology has recently offered great improvements in our research, overproduction of transport proteins being among others, a very important step forward. Hopefully, in the near future we will be able to purify and characterize the three membrane-bound components, thus allowing us to test biochemically the tenets of our models.

REFERENCES

Ames, G. F.-L. (1964). Uptake of amino acids by *Salmonella typhimurium*. *Arch. Biochem. Biophys.* **104,** 1–18.

Ames, G. F.-L. (1972). Components of histidine transport. Biological membranes. *ICN-UCLA Symp. Mol. Biol., 1st,* pp. 409–426.

Ames, G. F.-L. (1974). Isolation of transport mutants in bacteria. *In* "Methods in Enzymology: Biomembranes" (S. Fleischer, L. Packer, and R. W. Eastabrook, eds.), Vol. 32, pp. 843–849. Academic Press, New York.

Ames, G. F.-L., and Lever, J. (1970). Components of histidine transport: Histidine-binding proteins and *hisP* protein. *Proc. Natl. Acad. Sci. U.S.A.* **66,** 1096–1103.

Ames, G. F.-L., and Nikaido, K. (1978). Identification of a membrane protein as a histidine transport component in *S. typhimurium. Proc. Natl. Acad. Sci. U.S.A.* **75,** 5447–5451.

Ames, G. F.-L., and Roth, J. R. (1968). Histidine and aromatic permeases of *Salmonella typhimurium. J. Bacteriol.* **96,** 1742–1749.

Ames, G. F.-L., and Spudich, E. N. (1976). Protein-protein interaction in transport: Periplasmic histidine-binding protein J interacts with P protein. *Proc. Natl. Acad. Sci. U.S.A.* **73,** 1877–1881.

Ames, G. F.-L., Noel, K. D., Taber, H., Spudich, E. N., Nikaido, K , Afong, J., and Ardeshir, F. (1977). Fine-structure map of the histidine transport genes in *Salmonella typhimurium. J. Bacteriol.* **129,** 1289–1297.

Ardeshir, F., and Ames, G. F.-L. (1980). Cloning of the histidine transport genes from *Salmonella typhimurium* and characterization of an analogous transport system in *Escherichia coli. J. Supramol. Struct.* **13,** 117–130.

Bavoil, P., and Nikaido, H. (1981). Physical interaction between the phage λ receptor protein and the carrier-immobilized maltose-binding protein of *Escherichia coli. J. Biol. Chem.* **256,** 11385–11388.

Berger, E. A., and Heppel, L. A. (1974). Different mechanisms of energy coupling for the shock-sensitive and shock-resistant amino acid permeases of *Escherichia coli. J. Biol. Chem.* **249,** 7747–7755.

Boos, W. (1982). Aspects of maltose transport in *E. coli;* established facts and educated guesses. *Ann. Microbiol. (Inst. Pasteur).* **133A,** 145–151.

Brenner, M., and Ames, B. N. (1971). The histidine operon and its regulation. *In* "Metabolic Regulation" (H. J. Vogel, ed.), pp. 349–387. Academic Press, New York, New York.

Casadaban, M. H., and Cohen, S. N. (1979). Lactose genes fused to exogenous promoters in one step, using a Mu-*lac* bacteriophage: *In vivo* probe for transcriptional control sequences. *Proc. Natl. Acad. Sci. U.S.A.* **76,** 4530–4533.

Clark, A. F., and Hogg, R. W. (1981). High-affinity arabinose transport mutants of *E. coli:* Isolation and gene location. *J. Bacteriol.* **147,** 920–924.

Clement, J. M., and Hofnung, M. (1981). Gene sequence of the λ receptor, an outer membrane protein of *E. coli* K12. *Cell* **27,** 507–514.

Drummond, M., Clements, J., Merrick, M., and Dixon, R. (1983). Positive control and autogenous regulation of the *nifLA* promoter in *Klebsiella pneumoniae. Nature (London)* **301,** 302–307.

Furlong, C. E., and Schellenberg, G. D. (1980). Characterization of membrane proteins involved in transport. *In* "Microorganism and Nitorgen Sources" (J. W. Payne, ed.), pp. 89–123. Wiley, New York.

Gilliland, G. L., and Quiocho, F. A. (1981). Structure of the L-arabinose-binding protein from *Escherichia coli* at 2.4 Å resolution. *J. Mol. Biol.* **146,** 341–362.

Gilson, E., Higgins, C. F., Hofnung, M., Ames, G. F.-L., and Nikaido, H. (1982). Extensive homology between membrane-associated components of histidine and maltose transport systems of *Salmonella typhimurium* and *Excherichia coli. J. Biol. Chem.* **257,** 9915–9918.

Heppel, L. A. (1971). The concept of periplasmic enzymes. *In* "Structure and Function of Biological Membranes" (L. I. Rothfield, ed.), pp. 223–247. Academic Press, New York.

Higgins, C. F., and Ames, G. F.-L. (1981). Two periplasmic proteins which interact with a common membrane receptor show extensive homology: Complete nucleotide sequences. *Proc. Natl. Acad. Sci. U.S.A.* **78,** 6038–6042.

Higgins, C. F., Ames, G. F.-L., Barnes, W. M., Clement, J. M., and Hofnung, M. (1982a). A novel intercistronic regulatory element of prokaryotic operons. *Nature (London)* **298,** 760–762.

Higgins, C. F., Haag, P. D., Nikaido, K., Ardeshir, F., Garcia, G., and Ames, G. F.-L. (1982b). Complete nucleotide sequence and identification of membrane components of the histidine transport operon of *S. typhimurium. Nature (London)* **298,** 723–727.

Hobson, A., Weatherwax, R., and Ames, G. F.-L. (1984). ATP-binding sites in the membrane components of the histidine permease, a periplasmic transport system. *Proc. Natl. Acad. Sci. U.S.A.* **81,** 7333–7337.

Hogg, R. N. (1981). The amino acid sequence of the histidine-binding protein of *Salmonella typhimurium. J. Biol. Chem.* **256,** 1935–1939.

Hong, J.-H., Hunt, A. G., Masters, P. S., and Lieberman, M. A. (1979). Requirement of acetyl phosphate for the binding protein-dependent transport systems in *E. coli. Proc. Natl. Acad. Sci. U.S.A.* **76,** 1213–1217.

Hunt, A. G., and Hong, J. S. (1981). The reconstitution of binding protein-dependent active transport of glutamine in isolated membrane vesicles from *E. coli. J. Biol. Chem.* **256,** 11988–11991.

Krajewska-Grynkiewicz, K., Walczak, W., and Klopotowski, T. (1971). Mutants of *S. typhimurium* able to utilize D-histidine as a source of L-histidine. *J. Bacteriol.* **105,** 28–37.

Kustu, S. G., and Ames, G. F.-L. (1973). The *hisP* protein, a known histidine transport component in *S. typhimurium,* is also an arginine transport component. *J. Bacteriol.* **116,** 107–113.

Kustu, S. G., and Ames, G. F.-L. (1974). The histidine-binding protein J, a histidine transport component, has two different functional sites. *J. Biol. Chem.* **249,** 6976–6983.

Kustu, S. G., Chadwick McFarland, N., Hui, S. P., Esmon, B., and Ames, G. F.-L. (1979). Nitrogen control in *S. typhimurium:* Coregulation of synthesis of glutamine synthetase and amino acid transport systems. *J. Bacteriol.* **138,** 218–234.

Kyte, J., and Doolittle, R. F. (1982). A simple method for displaying the hydropathic character of a protein. *J. Mol. Biol.* **157,** 105–132.

Landick, R., Anderson, J. J., Mayo, M. M., Gunsalus, R. P., Mavromara, P., Daniels, C. J., and Oxender, D. L. (1980). Regulation of high-affinity leucine transport in *Escherichia coli. J. Supramol. Struct.* **14,** 527–537.

Magasanik, B. (1982). Genetic control of nitrogen assimilation in bacteria. *Annu. Rev. Genet.* **16,** 135–168.

Mahoney, W. C., Hogg, R. W., and Hermodson, M. A. (1981). The amino acid sequence of the D-galactose-binding protein from *Escherichia coli* B/r. *J. Biol. Chem.* **256,** 4350–4356.

Noel, K. D., and Ames, G. F.-L. (1978). Evidence for a common mechanism for the insertion of the Tn10 transposon and for the generation of Tn-10 stimulated deletions. *Mol. Gen. Genet.* **166,** 217–223.

O'Neill, M. C., Amass, K., and de Crombruggbe, B. (1981). Molecular model of the DNA interaction site for the cyclic AMP receptor protein. *Proc. Natl. Acad. Sci. U.S.A.* **78,** 2213–2217.

Oxender, D. L., Quay, S. C., and Anderson, J. J. (1980). Regulation of amino acid transport, *In* "Microorganisms and Nitrogen Sources" (J. W. Payne, ed.), pp. 153–169. Wiley, New York.

Pardee, A. B., and Prestidge, L. S. (1966). Cell-free activity of a sulfate binding site involved in active transport. *Proc. Natl. Acad. Sci. U.S.A.* **55,** 189–191.

Payne, J. (ed.) (1980). "Microorganisms and Nitrogen Sources." Wiley, New York.

Plate, C. A. (1979). Requirement for membrane potential in active transport of glutamine by *Escherichia coli*. *J. Bacteriol.* **137,** 221–225.

Robbins, A. R., Guzman, R., and Rotman, B. (1976). Roles of individual *mgl* gene products in the β-methylgalactoside transport system of *E. coli* K12. *J. Biol. Chem.* **251,** 3112–3116.

Robertson, D. E., Kroon, P. A., and Ho, C. (1977). Nuclear magnetic resonance and fluorescence studies of substrate-induced conformational changes of histidine-binding protein J of *Salmonella typhimurium*. *Biochemistry* **16,** 1443–1451.

Rosen, B. P. (1971). Basic amino acid transport in *Escherichia coli*. *J. Biol. Chem.* **246,** 3653–3662.

Rosen, B. P. (1978). "Bacterial Transport." Dekker, New York.

Shuman, H. A. (1982a). The maltose–maltodextrin transort system of *E. coli*. *Ann. Microbiol.* (*Inst. Pasteur*) **133A,** 153–159.

Shuman, H. A. (1982b). Active transport of maltose in *E. coli* K12. Role of the periplasmic maltose-binding protein and evidence for a substrate recognition site in the cytoplasmic membrane. *J. Biol. Chem.* **257,** 5455–5461.

Stern, M. J., Higgins, C. F., and Ames, G. F.-L. (1984a). Isolation and characterization of *lac* fusions to two nitrogen-regulated promoters. *Mol. Gen. Genet.* **195,** 219–227.

Stern, M. J., Ames, G. F.-L., Smith, N. H., Robinson, E. C., and Higgins, C. F. (1984b). Repetitive extragenic palindromic sequences: A major component of the bacterial genome. *Cell* **37,** 1015–1026.

Strange, P. G., and Koshland, D. E., Jr. (1976). Receptor interactions in a signaling system: Competition between ribose receptor and galactose receptor in the chemotaxis response. *Proc. Natl. Acad. Sci. U.S.A.* **73,** 762–766.

Yanofsky, C. (1981). Attenuation in the control of expression of bacterial operons. *Nature* (*London*) **289,** 751–758.

Zukin, R. S., Steinman, H. M., and Hirsch, R. E. (1984). Use of the distant reporter group method to study bacterial sensory receptors and transport proteins. *Microbiology* (in press).

Note Added in Proof

Recently we discovered that the P protein of the histidine transport system and the *malK* protein of maltose transport contain an ATP-binding site since they react with the light-activated affinity label 8-azido-ATP (Hobson *et al.*, 1984). Independently from us, a similar result was obtained for one of the components of the periplasmic oligopeptide transport system (C. F. Higgins, personal communication). These results strongly indicate that these proteins are involved in the energy-coupling step and that periplasmic systems probably use ATP, or a closely related compound, as an energy source. These results are in agreement with the hypothesis put forward by Berger and Heppel (1974) that substrate-level phosphorylation is the source of energy for periplasmic systems.

Chapter 7

A Study of Mutants of the Lactose Transport System of *Escherichia coli*

T. HASTINGS WILSON, DONNA SETO-YOUNG,*,[1] SYLVIE BEDU,*,[2] RESHA M. PUTZRATH,† AND BENNO MÜLLER-HILL‡*

**Department of Physiology*
Harvard Medical School
Boston, Massachusetts

†Environ Corporation
Washington, D.C.

and

‡Institut für Genetik der Universität zu Köln
Cologne, Federal Republic of Germany

I. INTRODUCTION

The lactose transport system of *Escherichia coli* belongs to a large class of membrane carriers found in all living cells (see Crane, 1977). This type of transport system, designated cation–substrate cotransport, carries out the net

[1]Present address: National Research Council of Canada, Division of Biological Sciences, Molecular Genetics Section, Ottawa, Ontario K1A OR6, Canada.
[2]Present address: U.E.R. Scientifique Marseille-Luminy, Physiologie Cellulaire, 13288 Marseille Cedex 9, France.

ISBN 0-12-153323-9

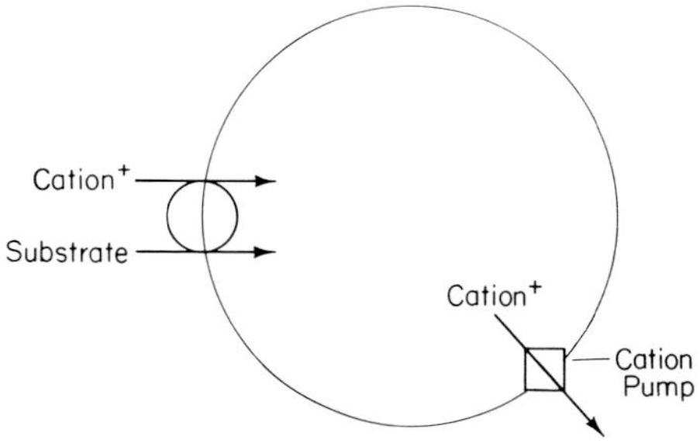

FIG. 1. Cation–substrate cotransport.

transfer of a substrate molecule plus a cation across the plasma membrane (Fig. 1). There is obligatory coupling between the movement of the substrate and cation. Thus the sudden addition of a high concentration of substrate to a cell with no cation gradient will result in an inward movement of the substrate down its concentration gradient, and a concomitant entry of cation against a concentration gradient. In the physiological situation the cation moves into the cell down its electrochemical potential difference and results in accumulation of substrate. In this situation the energy stored as a cation gradient drives the uphill transport of substrate.

This mechanism was proposed explicitly by Crane (1962), who studied the sodium-dependent glucose carrier of the mammalian small intestine. In 1963, Mitchell proposed that proton substrate cotransport may occur in microorganisms. He specifically suggested that proton lactose cotransport might be involved in the lactose carrier of *E. coli*. Direct experimental support for this suggestion came a few years later by West (1970). This was subsequently confirmed and extended by work in several different laboratories (Hirata *et al.*, 1974; Ramos *et al.*, 1976; Flagg and Wilson, 1977; Ahmed and Booth, 1981).

II. MUTANTS OF THE LACTOSE CARRIER

One approach to the study of the lactose carrier was the isolation and characterization of a variety of mutants. The first mutants of the lactose carrier were isolated by Monod and colleagues at the Pasteur Institute in Paris (Rickenberg *et al.*, 1956). These mutants were representatives of the commonest class, those with no detectable transport activity. Such mutants have been isolated and mapped by Malamy (1966), Langridge (1974), and Hobson *et al.* (1977).

A. *Km* Mutants

Another large class of mutants are those in which there is some physiological activity remaining. It is the purpose of this article to describe two subclasses of transport mutants in which there is altered transport activity.

One method for the isolation of mutants has been the thio-*o*-nitrophenylgalactoside (TONPG) method introduced by Müller-Hill *et al.* (1968). This sugar is toxic to induced cells, which accumulate sugar, within the cell. The toxicity is probably due to accumulation of the sugar, its leakage out of the cell, and its reaccumulation, which lead to a serious waste of metabolic energy and inhibition of growth (Wilson *et al.*, 1981). The technique involves growing cells with succinate as the carbon source, isopropylthiogalactoside as the inducer, and 2m*M* TONPG (Müller-Hill *et al.*, 1968; Smith and Sadler, 1971). Lactose-transport-negative cells grow perfectly normally whereas the lactose-positive cells fail to grow.

With this TONPG method a mutant (020) was isolated which is a representative of a class of mutants that have altered recognition of substrate (Flagg and Wilson, 1976). As shown in Table I, the K_m of the mutant for lactose was almost 10 times higher than that of the parent, whereas the V_{max} values were very similar for both cells. These cells fail to grow on agar plates containing 5 m*M* lactose, although growth is normal with 100 m*M* lactose. Furthermore, the cells grow normally on lactose if the cell is constitutive or if it is induced with isopropyl-β-thiogalactoside (IPTG). This suggests that the cell normally cannot accumulate sufficient intracellular lactose to induce the lactose operon.

This mutant also has very poor recognition for two other substrates: The entry rate of *o*-nitrophenylgalactoside (ONPG) was only 7% of that of the parent; thiomethylgalactoside (TMG) accumulation was 3% of that of the parent. The

TABLE I
PROPERTIES OF *Km* MUTANT (020)[a]

Lactose	K_m (m*M*)	V_{max} (μmol/g/min)
X71-15 (parent)	1.4	14.7
020	11.5	17.6
ONPG[b] (1 m*M*)	μmol OMP/min/g wet wt	
X71-15 (parent)	116	
020	8	
TMG (25 μ*M*)	Intracellular concentration in 30 min (μ*M*)	
X71-15 (parent)	1860	
020	60	

[a] Taken from Flagg and Wilson (1976).

[b] Values represent the thiodigalactoside (5 m*M*) inhibited rate of entry of ONPG.

rate of uptake of these two substrates was so poor that careful kinetic studies were difficult.

Mieschendahl *et al.* (1981) screened the collection of mutants in their laboratory (Hobson *et al.*, 1977) for possible examples of *Km* mutants. They discovered that 18 mutants grew well on 100 m*M* lactose but not on 5 m*M* lactose.

TABLE II
SUMMARY OF PROPERTIES OF *Km* MUTANTS[a]

Strain	Transport assays: ONPG entry (μmol ONP/min/mg dry wt)	Transport assays: TMG uptake (in/out)	Sugar fermentation (on plates)[b]: Lactose − IPTG	Lactose + IPTG	Melibiose − IPTG	Melibiose + IPTG
DP90/F′lac (parent)	1.7	66	Red	Red	White	Red
020	0.04	1.7	White	Red	White	White
AG47	0	2.5	Red center[c]	Red	White	Red
AA22	0	2.2	Red center[c]	Red	Red center	Red
MAB20	0	2.1	White	Red	White	Red
AN14	0	1.5	White	Red	White	Red
AV38	0	1.6	Red center	Red	White	Red
AE43	0	1.9	White	Red	White	Red
AC43	0	4.8	Red center	Red	White	White
AV40	0	3.3	White	Red	White	White
AJ36	0	2.2	White	Red	White	Red
AE10	0	2.2	White	Red	White	White
AJ33	0	5.5	White	Red	White	White
AD47	0.02	1.4	Red center	Red	White	Red
NP 4	0.04	1.5	White	Red	White	Red center
AG38	0.02	2.0	White	Red center	White	White
AE41	0	1.5	Red center	Red	White	Red
AF34	0	2.5	White	Red	White	Red
MAA36	0	1	White	Red	White	Red
MNB 7	0	2	White	Red	White	Red center

[a] The parental (*DP90/F′lac*) genotype is $\Delta lac/F'I^{Q}Z^{+}Y^{+}$. The mutants are all isolated from this parent (except 020 which was originally a chromosomal *Y* gene mutant which was placed on an $F'I^{Q}Z^{+}$ in the DP90 background). ONPG (1 m*M*) entry was measured in potassium phosphate buffer, pH 7, at 20°C. TMG (0.1 m*M*) accumulation was measured in the same buffer at 20°C. Data are expressed as the ratio of the concentration inside divided by the concentration outside at 5 min. MacConkey indicator plates contained 1% sugar (lactose or melibiose) with or without 1 m*M* isopropylthiogalactoside and were incubated for 18 hr at 37°C.

[b] Red indicates fermentation, white indicates the absence of fermentation, red center indicates weak fermentation.

[c] When tested with 0.2% lactose in the plate, these cells gave white clones while the parental cell was red.

These mutants plus 020 were tested for transport and for fermentation of lactose and melibiose (Table II).

All mutants showed enhanced lactose fermentation in the presence of IPTG. Most mutants showed a similar stimulation of fermentation of melibiose by IPTG; however six failed to show positive fermentation with or without inducer. All mutants in Table II showed a marked defect in the transport of ONPG and TMG.

Another type of mutant with an altered sugar recognition site was isolated by Shuman and Beckwith (1979). They isolated a *lacY* mutant that was able to recognize maltose, a sugar normally not transported by the lactose carrier. This mutant showed a somewhat greater affinity for ONPG and TMG than the parent and a normal V_{max} value for the two substrates. Mieschendahl *et al.* (1981) screened the 18 *Km* mutants and found that one of them (AJ33) could utilize maltose via the lactose carrier.

B. Energy-Uncoupled Mutants

An extremely rare type of mutant (ML-308-22) was isolated by the TONPG method described above (Wong *et al.*, 1970). Like the *Km* mutants, this cell showed a defect in accumulation of substrates such as TMG. An unexpected finding was that entry of ONPG was faster into the mutant than into the parent. At room temperature, its V_{max} was 200% that of the parent (its K_m was four times higher than that of the parent). At 10°C, the V_{max} of the mutant was 400% of the parent (Wilson, 1978). The high rate of entry was not due to general leakiness of the cell because the ONPG entry was blocked by competitive inhibitors of the carrier. In addition, the transport of other sugars was found to be normal in this cell.

Since the accumulation of sugar by this carrier is driven by proton entry, it was postulated that perhaps proton recognition was abnormal and protons do not enter the cell. Proton entry associated with lactose entry was measured by adding a galactoside to energy-depleted cells and measuring proton uptake as an alkaline pH shift in the external medium.

Cells were first energy depleted by incubating them under nitrogen to stop the respiratory proton pumps. Cells were incubated in a very lightly buffered solution and the pH monitored continuously (West and Wilson, 1973). When TMG was added to the parental cell, there was a marked alkalinization of the external medium, as protons entered the cell (Fig. 2). The pH change was temporary, however, because sugar equilibration across the membrane resulted in no further net sugar or proton uptake and allowed the gradual dissipation of the pH gradient over a period of 5–10 min. In the lower panel of Fig. 2 is shown a similar experiment with the mutant. Addition of sugar resulted in very little proton

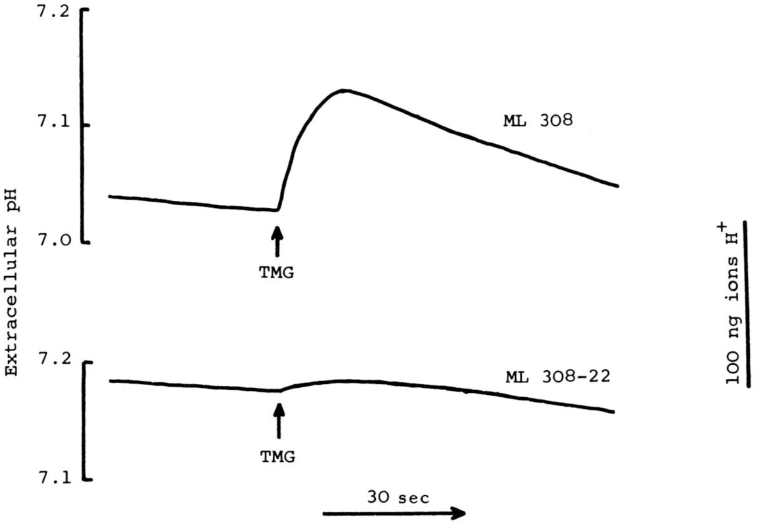

FIG. 2. Extracellular pH changes on adding TMG to anaerobic suspensions of cells. To the 2 ml of suspension of ML308 or ML308-22 was added 10 μl of anaerobic 1 *M* TMG at the arrow. From West and Wilson (1973).

movement. We concluded that although sugar could be recognized by the carrier, entry occurred in the absence of proton movement.

The original cell of this type was isolated in the ML strain, which is extremely difficult to manipulate genetically. Another similar mutant (X71-54) was isolated in the K12 strain (Wilson *et al.*, 1970). This cell showed the same phenotype as ML-22. It showed a defect in accumulation of TMG while the ONPG entry (V_{max}) was about 200% of the parent.

1. Reconstitution Studies

It was desirable to obtain data on the transport of the natural substrate, lactose. In order to avoid artifacts due to the metabolism of the sugar, it was decided to study transport in the carrier reconstituted in proteoliposomes.

The reconstitution technique is indicated diagrammatically in Fig. 3 (Newman and Wilson, 1980). Cell membranes were prepared by disrupting the cells with a French pressure cell. This caused the cells to rupture, with formation of small membrane fragments. These fragments were then centrifuged and washed. The lactose carrier was next extracted from the membrane with octylglucoside in the presence of added *E. coli* phospholipid. The unextracted membranes were then removed by centrifugation. At this stage the supernatant fluid contained micelles of phospholipid detergent and protein. To the supernatant was added additional

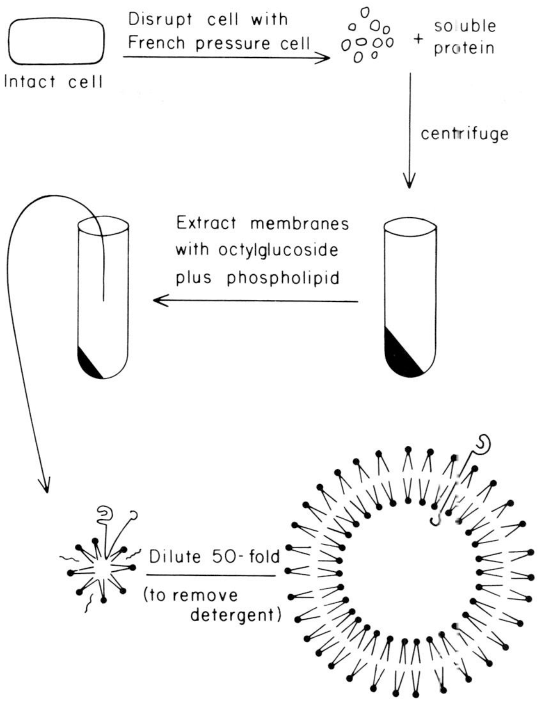

FIG. 3. Technique of reconstitution for the preparation of proteoliposomes.

phospholipid, and the mixture was diluted 50-fold according to the procedure suggested by Racker *et al.* (1979). Most of the mycellar phospholipid was converted to unilamellar liposomes (100 nm in diameter) containing carrier protein. These proteoliposomes were centrifuged and resuspended in the desired medium.

One transport assay was the incubation of these proteoliposomes with radioactive lactose. Sugar entered the proteoliposomes rapidly (Fig. 4) and equilibrated across the membrane. In order to induce accumulation the proteoliposomes were preloaded with 20 m*M* nonradioactive lactose, centrifuged, and resuspended in 0.4 m*M* [^{14}C]lactose. There was a 10-fold accumulation of radioactive lactose. The radioactive molecules that entered the proteoliposomes were trapped inside since the cold molecules occupied the sites for exit. This accumulation was only temporary, however, as the exit for cold molecules of lactose progressively reduced the competitive inhibition of exit of hot molecules. Ultimately both hot and cold molecules equilibrated across the membrane. In these experiments it was the accumulation aspect of these counterflow curves that was studied. The addition of a potent competitive inhibitor (*p*-nitrophenyl-α-galactoside) to the outside blocked entry of hot lactose.

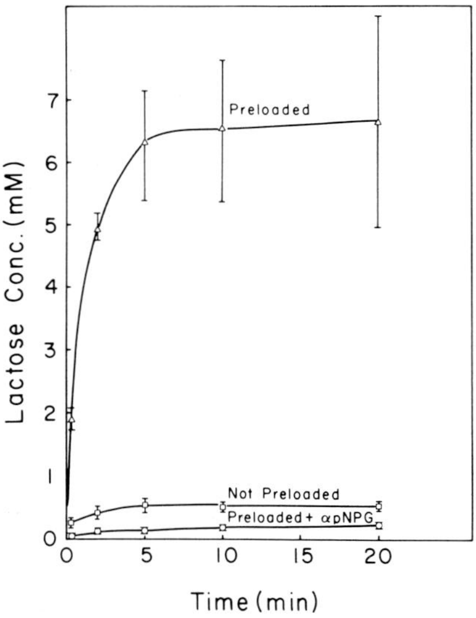

FIG. 4. Lactose uptake in lactose-preloaded and non-preloaded proteoliposomes. The lactose-pre-loaded (20 m*M*) or non-preloaded proteoliposomes were prepared from the extract of membrane vesicles of *E. coli* X71/F′W3747 (*lac* $IZ^+Y^+/F'I^+Z^+Y^+$). The final external concentration of [^{14}C]lactose was 0.4 m*M* with or without 10 m*M* α-*p*NPG (pH 6). △——△, Lactose-preloaded proteoliposomes; □——□, non-preloaded proteoliposomes; ○——○, lactose-preloaded proteoliposomes with 10 m*M* α-*p*NPG in the assay medium. Data given represent the mean values of three experiments. The error bar represents the standard error. From Seto-Young *et al.* (1984).

The counterflow observed with proteoliposomes prepared from the mutant carrier vs those from parental strain was compared. The initial rate of entry was more rapid in the mutant, and the steady-state accumulation was slightly higher in the mutant (Fig. 5). Kinetic studies of the counterflow phenomenon indicated that the K_m for the mutant was 30% that of the parent, whereas the V_{max} values for both were similar (Seto-Young *et al.*, 1984).

The lactose uptake driven by a protonmotive force (pH gradient plus membrane potential) was then measured in parent and mutant. Proteoliposomes preloaded with potassium phosphate, pH 7.5, were incubated in an external medium of sodium phosphate, pH 6, in the presence of valinomycin. Exit of K^+ on the

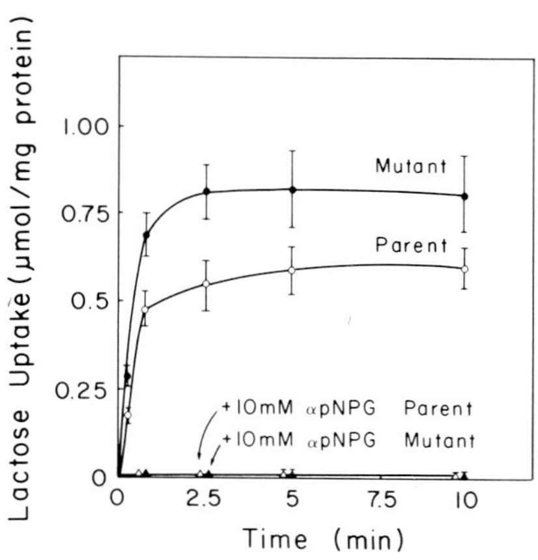

FIG. 5. Counterflow in proteoliposomes from parent and mutant. Lactose-preloaded proteoliposomes were prepared from the extract of membrane vesicles of parental cells (X71/F′W3747) and mutant cells [54/F′5441 (*lac* $I^-Z^+Y^{UN}/F'lac\ I^+Z^+Y^{UN}$)]. The transport assay was carried out at pH 6 with 0.4 m*M* [^{14}C]lactose. ○——○ and △——△; Parent; ●——● and ▲——▲, mutant. Data given represent the mean values of three experiments (with standard error). From Seto-Young *et al.* (1984).

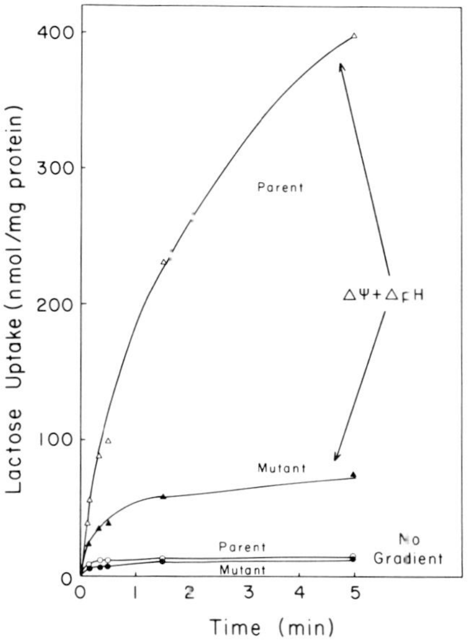

FIG. 6. pH gradient- and membrane potential-driven lactose uptake in proteoliposomes from parent and mutant. Proteoliposomes were preloaded with 100 m*M* potassium phosphate plus 25 m*M* MES, pH 7.5. Valinomycin was added to the concentrated suspension of proteoliposomes to give a concentration of 19 μ*M*. Proteoliposomes were diluted 100-fold into an assay medium containing 100 m*M* potassium phosphate or sodium phosphate, 25 m*M* MES, pH 7.5 or 6, and 0.2 m*M* lactose. The open symbols represent the parent and the closed symbols represent the mutant. △——△ and ▲——▲, pH Gradient and membrane potential (inside pH 7.5 and K^+; outside pH 6 and Na^+); ○——○ and ●——●, no pH gradient or membrane potential (inside pH 7.5 and K^+; outside pH 7.5 and K^+). From Seto-Young *et al.* (1984).

valinomycin resulted in a K^+ diffusion potential inside negative. Proteoliposomes prepared from parental lactose carrier accumulated 30-fold, whereas the accumulation of the mutant was approximately 5-fold above that of the external medium (Fig. 6). A similar defect in transport was shown by the mutant when the driving force was pH gradient alone or membrane potential alone (Seto-Young *et al.,* 1984).

Thus the mutant shows the unusual property of a membrane carrier that recognizes the lactose slightly better than does the parent, but shows a defect in the movement against a concentration gradient. Since it is known that this mutant showed reduced proton uptake in response to addition of galactoside (West and Wilson, 1973), it appeared that the mutant was capable of the uptake of lactose in the absence of protons (or with reduced number of protons). It appears reasonable to postulate that in the mutant the sugar recognition site is intact, whereas the proton recognition site is abnormal.

2. BINDING STUDIES

One further experiment of a different type suggests that the sugar recognition site of the mutant carrier has a higher affinity for one of its substrates than normal. The binding of *p*-[^{3}H]nitrophenyl-α-galactoside (α-*p*NPG) to membranes of mutant and parental cells was compared [by the method of Kennedy *et al.* (1974)]. Membranes of the mutant incubated with 12 μ*M* α-*p*NPG bound more than twice as much sugar as membranes from parental cells (Table III).

TABLE III
α-*p*NPG BINDING TO MEMBRANES OF PARENT AND MUTANT CELLS[a]

	pmol α-*p*NPG bound/mg membrane protein at pH			
Cell	7.0	6.0	5.5	5.0
Parent X71/X71	84	84	63	60
Mutant 54/54	210	264	254	143

[a] The genotype of X71/X71 is $I^-Z^+Y^+A^-/F'I^-Z^+Y^+A^-$ and 54/54 is $I^-Z^+Y^{UN}A^-/F'I^-Z^+Y^{UN}A^-$. Binding of α-*p*NPG by membranes from the parental and mutant cells was determined by the method of Kennedy *et al.* (1974). Membranes were exposed to 12 μ*M* [^{3}H]α-*p*NPG with or without 8 m*M* cold α-*p*NPG. The value obtained with 8 m*M* cold α-*p*NPG plus hot α-*p*NPG was subtracted from that with hot α-*p*NPG alone. Membranes were centrifuged down at 40,000 *g* for 1 hr and pellet counted. Data represent the mean values of two independent experiments.

This is consistent with the hypothesis that the mutant carrier possesses a higher affinity for α-*p*NPG than does the parent.

3. Y^{f4} MUTATION

There is another mutant of this class which is of interest. This mutant, designated Y^{f4}, has been studied by Victor Fried (1977, 1981). It will grow on lactose only in the presence of IPTG. It shows normal entry of ONPG but no accumulation of TMG. Previous genetic studies (see Fried, 1977) indicated that this mutant possessed two separate defects, one toward the terminal end of the *Y* gene

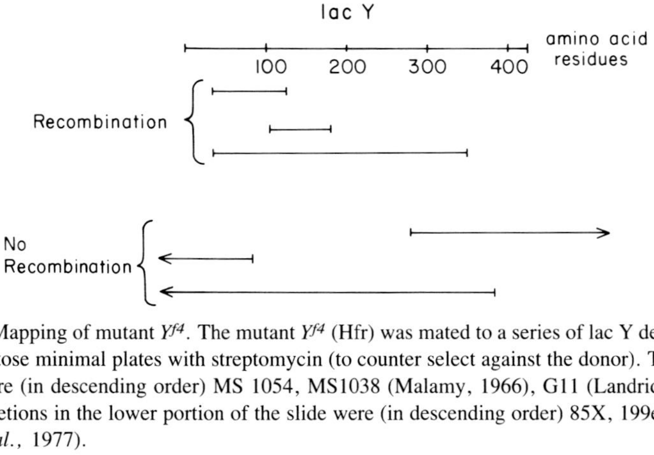

FIG. 7. Mapping of mutant Y^{f4}. The mutant Y^{f4} (Hfr) was mated to a series of lac Y deletions and plates on lactose minimal plates with streptomycin (to counter select against the donor). The internal deletions were (in descending order) MS 1054, MS1038 (Malamy, 1966), G11 (Landridge, 1974). The long deletions in the lower portion of the slide were (in descending order) 85X, 199e, and 202a (Hobson *et al.*, 1977).

and a second near the C-terminal region of the *Y* gene. This conclusion has been confirmed by mapping studies shown in Fig. 7. Recombination was observed with three different internal deletions. On the other hand, no recombination was observed when mated to deletions of the N-terminal region or a deletion of the C terminus. These data are consistent with the view that there is one defect at each end of the *Y* gene. In an attempt to remove the N-terminal mutation, the mutant was crossed with a C-terminal deletion (85X) from the Müller-Hill collection (Hobson *et al.*, 1977). After counterselecting against the donor with streptomycin, it was possible to obtain a recombinant that grew slowly on lactose. This proved to have only one mutation—at the C-terminal region. This cell showed ONPG entry of 50% of normal but no TMG accumulation. It grew on lactose in the presence of IPTG.

III. DISCUSSION

These findings should be considered in context of important new knowledge of the molecular biology of the lactose carrier. The *lacY* gene of *E. coli* was first cloned on the plasmid PBR 322 by Teather *et al.* (1978, 1980). This was an extremely important advance, as it allowed the DNA sequencing (Büchel *et al.*, 1980) and the purification of the protein (Newman *et al* , 1981; Wright *et al.*, 1982). The DNA sequencing allowed the assignment of an amino acid sequence. The N-terminal region was confirmed by chemical analysis of the first several N-terminal amino acids by Ehring *et al.* (1980). Furthermore, this group synthesized the lactose carrier *in vitro* and demonstrated that the N-terminal sequence was similar to that predicted by the DNA sequence. Since the mature lactose carrier *in vivo* possesses the same terminal sequence, processing the N terminal was excluded. Experiments of Seto-Young *et al.* (1984) have shown that, when the purified lac carrier is exposed to carboxypeptidase Y, alanine and valine are released. Since these are the two carboxy-terminal amino acids predicted by the DNA sequence, it appears likely that there is no processing at the C-terminal end of the molecule.

As Büchel *et al.* (1980) pointed out in their DNA sequencing paper, the

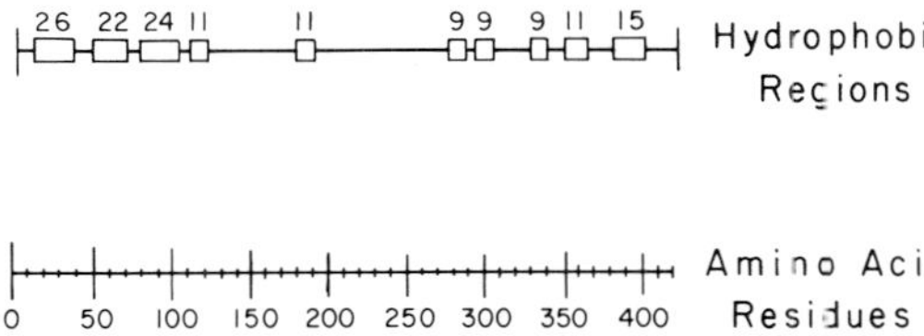

FIG. 8. Hydrophobic regions of the lactose carrier. The squares represent regions of consecutive hydrophobic residues. From Mieschendahl *et al.* (1981).

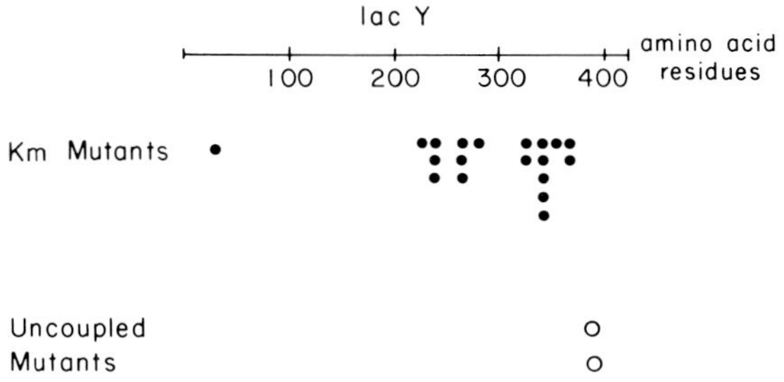

FIG. 9. Mapping of *Km* and "uncoupled" mutants. From Hobson *et al.* (1977) and Mieschendahl *et al.* (1981).

protein is very hydrophobic: 70% of the amino acid residues are hydrophobic. Mieschendahl *et al.* constructed a simple figure to illustrate the groupings of hydrophobic residues (Fig. 8). These authors suggest that the lac carrier protein passes in and out of the membrane.

The map location of defects in the *Km* and energy-uncoupled mutants is of interest in the structural studies of the transport protein (Fig. 9). Eighteen of 19 mutants map in the second half of the DNA molecule (Hobson *et al.*, 1977; Mieschendahl *et al.*, 1981). Mutant 020, for which we have the most physiological data, maps with several other *Km* mutants toward the C-terminal end of the molecule. Because most of the *Km* mutants map in the second half of the molecule, it was suggested by Mieschendahl *et al.* (1981) that the sugar recognition site resided in the second half of the molecule, probably a channel formed by several helixes passing through the membrane. One such model (Beyreuther, 1982) shows 14 segments that penetrate the membrane. A somewhat analogous conclusion was reached by Foster *et al.* (1983) on the basis of physical measurements of the α-helical structure of the protein. The SH group essential for lactose carrier activity was identified by Beyreuther *et al.* (1981) as the cysteine residue at position 148.

The two uncoupled mutants both mapped to the extreme C-terminal end of the *Y* gene (Fig. 9). On the basis of these data it is proposed that the cation recognition site may involve an amino acid residue somewhere in the C-terminal 30 amino acids of the polypeptide chain.

REFERENCES

Ahmed, S., and Booth, I. R. (1981). Quantitative measurements of the protonmotive force and its relation to steady state lactose accumulation in *Escherichia coli*. *Biochem. J.* **200,** 573–581.

Beyreuther, K., [quoted by Overath, P., and Wright, J. K. (1982). Lactose permease and the molecular biology of transport. *Hoppe-Seyler's Z. Physiol. Chem.* **363,** 1409–1414.]

Beyreuther, K., Bieseler, B., Ehring, R., and Müller-Hill, B. (1981). Identification of internal residues of lactose permease of *Escherichia coli* by radiolabel sequencing of peptide mixtures, *In* "Methods in Protein Sequence Analysis" (M. Elzinga, ed.), pp. 139–148. Humana Press, Clifton, New Jersey.

Büchel, D. E., Gronenborn, B., and Müller-Hill, B. (1980). Sequence of the lactose permease gene. *Nature (London)* **283,** 541–545.

Crane, R. K. (1962). Hypothesis for mechanism of intestinal active transport of sugars. *Fed. Proc.* **21,** 891–895.

Crane, R. K. (1977). The gradient hypothesis and other models of carrier-mediated active transport, *Rev. Physiol. Biochem. Pharmacol.* **78,** 99–159.

Ehring, R., Beyreuther, K., Wright, J. K., and Overath, P. (1980). *In vitro* and *in vivo* products of *E. coli* lactose permease gene are identical. *Nature (London)* **283,** 537–540.

Flagg, J. L., and Wilson, T. H. (1976). *Lac*Y mutant of *Escherichia coli* with altered physiology of lactose induction. *J. Bacteriol.* **128,** 701–707.

Flagg, J. L., and Wilson, T. H. (1977). A protonmotive force as the source of energy for galactoside transport in energy depleted *Escherichia coli. J. Membr. Biol.* **31,** 233–255.

Foster, D., Boublik, M., and Kaback, H. R. (1983). Structure of the *lac* carrier protein of *Escherichia coli. J. Biol. Chem.* **258,** 31–34.

Fried, V. A. (1977). A novel mutant of the *lac* transport system of *E. coli. J. Mol. Biol.* **114,** 477–490.

Fried, V. A. (1981). Membrane biogenesis: Evidence that a soluble chimeric polypeptide can serve as a precursor of a mutant *lac* permease in *Escherichia coli. J. Biol. Chem.* **256,** 244–252.

Hirata, H., Altendorf, K., and Harold, F. M. (1974). Energy coupling in membrane vesicles of *Escherichia coli. J. Biol. Chem.* **249,** 2939–2945.

Hobson, A. C., Gho, D., and Müller-Hill, B. (1977). Isolation, genetic analysis, and characterization of *Escherichia coli* mutants with defects in the *lac*Y gene. *J. Bacteriol.* **131,** 830–838.

Kennedy, E. P., Rumley, M. K., and Armstrong, J. B. (1974). Direct measurement of the binding of labeled sugars to the lactose permease M Protein. *J. Biol. Chem.* **249,** 33–37.

Langridge, J. (1974). Characterization and intragenic position of mutations in the gene for galactoside permease of *Escherichia coli. Aust. J. Biol. Sci.* **27,** 331–340.

Malamy, M. H. (1966). Frameshift mutations in the lactose operon of E. coli. Cold Spring Harbor Symp. **31,** 189–201.

Mieschendahl, M., Büchel, D., Bocklage, H., and Müller-Hill, B. (1981). Mutations in the lacY gene of *Escherichia coli* define functional organization of lactose permease. *Proc. Natl. Acad. Sci. U.S.A.* **78,** 7652–7656.

Mitchell, P. (1963). Molecule, group and electron translocation through natural membranes. *Biochem. Soc. Symp.* **22,** 142–168.

Müller-Hill, B., Crapo, L., and Gilbert, W. (1968). Mutants that make more lac repressor. *Proc. Natl. Acad. Sci. U.S.A.* **59,** 1259–1272.

Newman, M. J., and Wilson, T. H. (1980). Solubilization and reconstitution of the lactose transport system from *Escherichia coli. J. Biol. Chem.* **255,** 10583–10586.

Newman, M. J., Foster, D. L., Wilson, T. H., and Kaback, H. R. (1981). Purification and reconstitution of functional lactose carrier from *Escherichia coli. J. Biol. Chem.* **256,** 11804–11808.

Racker, E., Violand, B., O'Neal, S., Alfonzo, M., and Telford, J. (1979). Reconstitution, a way of biochemical research; some new approaches to membrane-bound enzymes. *Arch. Biochem. Biophys.* **198,** 470–477.

Ramos, S., Schuldiner, S., and Kaback, H. R. (1976). The electrochemical gradient of protons and its relationship to active transport in *Escherichia coli* membrane vesicles. *Proc. Natl. Acad. Sci. U.S.A.* **73,** 1892–1896.

Rickenberg, H. V., Cohen, G. N., Butlin, G., and Monod, J. (1956). La galactoside-permease d'*Escherichia coli. Ann. Inst. Pasteur (Paris)* **91,** 829–857.

Seto-Young, D., Bedu, S., and Wilson, T. H. (1984). Transport by reconstituted lactose carrier from parental and mutant strains of *Escherichia coli. J. Membr. Biol.* **79,** 185–193.

Shuman, H. A., and Beckwith, J. (1979). *Escherichia coli* K-12 mutants that allow transport of maltose via the β-galactoside transport system. *J. Bacteriol.* **137,** 365–373.

Smith, T. F., and Sadler, J. R. (1971). The nature of lactose operator constitutive mutations. *J. Mol. Biol.* **59,** 273–305.

Teather, R. M., Müller-Hill, B., Abrutsch, U., Aichele, G., and Overath, P. (1978). Amplification of the lactose carrier protein in *Escherichia coli* using a plasmid vector. *Mol. Gen. Genet.* **159,** 239–248.

Teather, R. M., Bramhall, J., Riede, I., Wright, J. K., Furst, M., Aichele, G., Wilhelm, U., and Overath, P. (1980). Lactose carrier protein of *Escherichia coli:* Structure and expression of plasmids carrying the Y gene of the *lac* operon. *Eur. J. Biochem.* **108,** 223–231.

West, I. (1970). Lactose transport coupled to proton movements in *Escherichia coli. Biochem. Biophys. Res. Commun.* **41,** 655–661.

West, I. C., and Wilson, T. H. (1973). Galactoside transport dissociated from proton movement in mutants of *Escherichia coli. Biochem. Biophys. Res. Commun.* **50,** 551–558.

Wilson, D. M., Putzrath, R., and Wilson, T. H. (1981). Inhibition of growth of *Escherichia coli* by lactose and other galactosides. *Biochim. Biophys. Acta* **649,** 377–384.

Wilson, T. H. (1978). Lactose transport in *Escherichia coli. In* "Physiology of Membrane Disorders" (T. E. Andreoli, J. F. Hoffman, and D. P. Fanestil, eds.), pp. 459–476. Plenum, New York.

Wilson, T. H., Kusch, M., and Kashket, E. R. (1970). A mutant in *Escherichia coli* energy uncoupled for lactose transport; a defect in the lactose-operon. *Biochem. Biophys. Res. Commun.* **40,** 1407–1414.

Wong, P. T. S., Kashket, E. R., and Wilson, T. H. (1970). Energy coupling in the lactose transport system of *Escherichia coli. Proc. Natl. Acad. Sci. U.S.A.* **65,** 63–69.

Wright, J. K., Schwarz, H., Straub, E., Overath, P., Bieseler, B., and Beyreuther, K. (1982). Lactose carrier protein of *Escherichia coli:* Reconstitution of galactoside binding and countertransport. *Eur. J. Biochem.* **124,** 545–552.

Chapter 8

The Proton-ATPase of *Escherichia coli*

A. E. SENIOR

Department of Biochemistry
University of Rochester Medical Center
Rochester, New York

I. UBIQUITY OF PROTON-ATPases

Proton-ATPases are electrogenic proton pumps that generate transmembrane electrochemical gradients of protons ($\Delta\bar{\mu}H^+$) composed of component pH gradients (ΔpH) and membrane potentials ($\Delta\psi$). The "F_1F_0-type" of proton-ATPase occurs in the plasma membrane of bacteria and is an essential component of bacterial energy transductions (Simoni and Postma, 1975; Haddock and Jones, 1977). Figure 1 shows the two main ways in which proton potentials ($\Delta\bar{\mu}H^+$) may be generated in bacteria, i.e., by the plasma membrane respiratory chain proton pumps ("electron transfer") or by the proton-ATPase ("ATP hydrolysis"). The protons are pumped outward in each case, so that the interior (cytoplasm) becomes negative. A major function of the $\Delta\bar{\mu}H^+$ in all bacteria is to drive the uptake of essential nutrients (e.g., sugars, amino acids), which

ISBN 0-12-153323-9

occurs by symport with H^+ and is mediated by specific transport proteins in the plasma membrane. In bacteria that lack respiratory chain H^+ pumps, or under anaerobic conditions, the proton-ATPase (using glycolytic ATP) may generate the $\Delta\bar{\mu}H^+$. In aerobic bacteria, under conditions where the respiratory chain H^+ pumps are generating $\Delta\bar{\mu}H^+$, the proton-ATPase will run "backwards," i.e., protons will enter the cell through the proton-ATPase, driving ATP synthesis (oxidative phosphorylation) (see Fig. 1). Two other functions that require the energy of the proton potential in bacteria are flagellar motion and protein secretion (not shown in Fig. 1).

F_1F_0-type proton-ATPases occur also in the inner membrane of mitochondria and the thylakoid membrane of chloroplasts where they catalyze oxidative phosphorylation and photophosphorylation, respectively, coupled to proton potentials generated by respiration and light capture (Senior, 1979; Shavit, 1980).

A proton-ATPase occurs in the plasma membranes of yeasts and *Neurospora crassa* which is of a type clearly distinct in structure from the F_1F_0 type (Goffeau and Slayman, 1981). The proton-pumping ATPase of gastric mucosa (Sachs *et al.*, 1982) is also clearly distinct from the F_1F_0 type and shows some structural similarity to the yeast and *N. crassa* enzymes. Proton-ATPases have been reported in lysosomes (Harikumar and Reeves, 1983), sperm acrosome (Working and Meizel, 1981), intracellular vacuoles of yeast (Kakinuma *et al.*, 1981), chromaffin granules of adrenal gland (Apps, 1982), clathrin-coated vesicles from brain (Stone *et al.*, 1983), turtle urinary bladder (Gluck *et al.*, 1982), and plant cell plasma membranes (Vara and Serrano, 1982). The information to date is not extensive enough to allow us to classify the latter enzymes with confidence. However, it assuredly seems that proton-ATPases are ubiquitous and have three major physiological roles: (1) In mitochondria, chloroplasts, or bacteria, ATP synthesis is catalyzed using $\Delta\bar{\mu}H^+$ to drive the reaction. (2) $\Delta\bar{\mu}H^+$ is generated by ATP hydrolysis and is used to drive transport reactions. (Examples are the yeast and *Neurospora* plasma membrane enzymes, the chromaffin granule en-

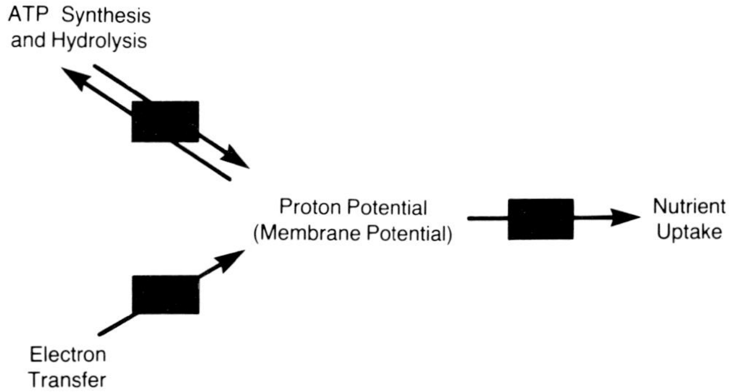

FIG. 1. Generation and utilization of proton potentials in bacteria.

zyme, and the bacterial F_1F_0 type under anaerobic conditions or in organisms lacking respiratory chain H^+ pumps.) (3) Protons are pumped to sustain a required (low) pH milieu. (Examples are the enzymes of gastric mucosa, lysosomes, and clathrin-coated vesicles.)

This article will describe in detail the F_1F_0-type proton-ATPase from *Escherichia coli*. Genetic manipulation has greatly facilitated the study of this enzyme, such that it is now the best characterized proton-ATPase. I shall try to show how the application of genetic techniques has contributed to understanding the structure, the mechanism of proton conduction, the mechanism of catalysis, and the *in vivo* assembly of the enzyme.

II. RESOLUTION AND RECONSTITUTION OF THE F_1 SECTOR AND THE MEMBRANE SECTOR (F_0)

Figure 2 shows an experiment that resolves *E. coli* proton-ATPase into two sectors and then reconstitutes them. Washing the purified plasma membranes in buffer containing EDTA at very low ionic strength releases one part of the enzyme, the F_1 sector, which may be subsequently purified as a soluble ATPase,

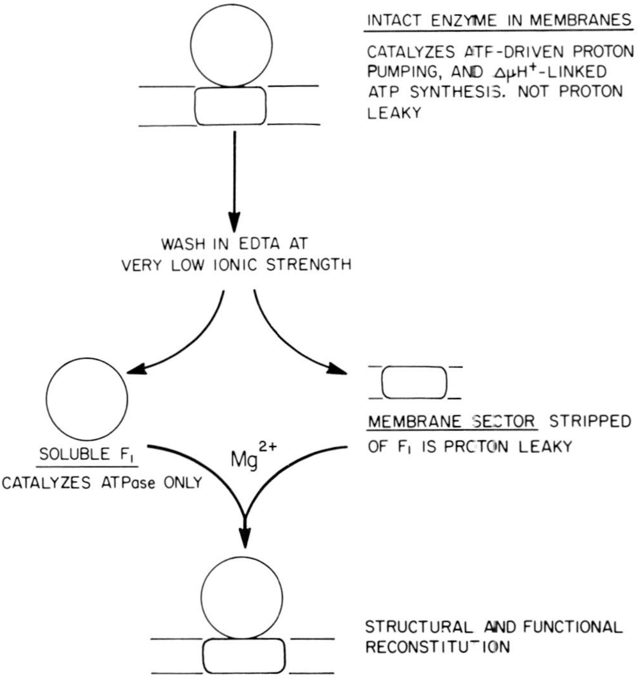

FIG. 2. Resolution–reconstitution of the H^+-ATPase of *E. coli*.

and leaves embedded in the membrane the membrane sector (also called "F_0"), which forms the proton channel. Reconstitution is achieved when the F_1 and F_0 are incubated together in Mg-containing buffer. This experiment is based on the original studies by Pullman, Penefsky, and Racker of the mitochondrial proton-ATPase and by Abrams of *Streptococcus faecalis* proton-ATPase, and together with later complementary studies (reviewed in Fillingame, 1980, 1981) shows that in the F_1F_0-type proton-ATPases a catalytic unit (F_1), on which ATP synthesis and hydrolysis occur, is integrated with a proton channel (F_0). The F_1 sector is extrinsic to the membrane and is composed of water-soluble subunits. The F_0 is intrinsic to the membrane and is composed of largely hydrophobic subunits.

III. GENES AND SUBUNITS OF THE *E. coli* PROTON-ATPase

The elucidation of the genetics of the enzyme was largely the work of Gibson, Cox, Downie, and co-workers (reviewed in Downie *et al.*, 1979; Gibson, 1983). These workers obtained a large number of mutants (which they designated as *unc* for *unc*oupled), developed a genetic complementation test system, established several gene–subunit relationships, and showed that the structural genes are clustered in an operon on the *E. coli* chromosome. The DNA sequence of this entire operon was then determined by three different groups (reviewed in Senior and Wise, 1983), and it has been possible to make predictions of secondary and tertiary structure of the subunits from the amino acid sequences and to predict the positions of nucleotide-binding sites (e.g., Mabuchi *et al.*, 1981; Walker *et al.*, 1982a).

Biochemical studies had suggested that F_1 contained five types of subunit and that F_0 contained three types (Fillingame, 1981). Figure 3 indicates the locations of the subunits in the enzyme together with their approximate molecular weights and corresponding genes. The stoichiometry of subunits in F_1 is $\alpha_3\beta_3\gamma\delta\epsilon$ (Senior and Wise, 1983); the stoichiometry of F_0 subunits is not yet settled, but it is known there are multiple copies (6 to 10) of the uncE protein.

The *unc* operon is shown in Fig. 4, and it is of interest from several points of

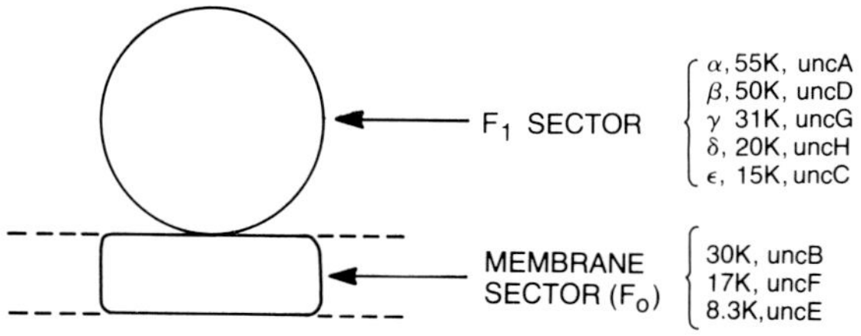

FIG. 3. Subunits and structural genes of the *E. coli* H^+-ATPase.

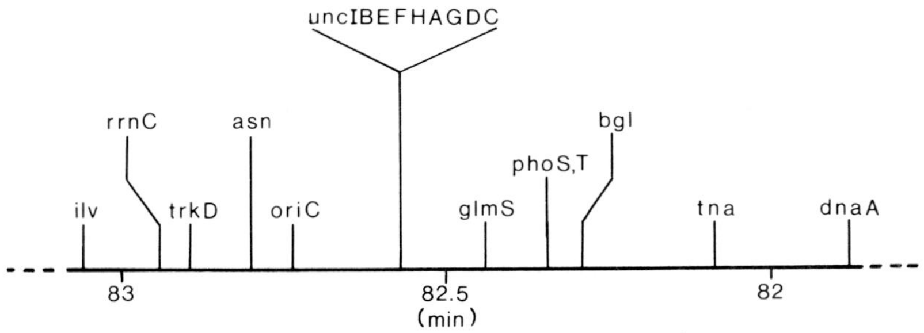

FIG. 4. Location and order of genes in the *E. coli* H^+-ATPase. (From Senior and Wise, 1983, with permission.)

view. The first gene (*uncI*) is not a structural gene, apparently. It was discovered only when the DNA was sequenced as an open reading frame which would code for a basic, hydrophobic protein. However, mutations in this gene did not affect synthesis, assembly, or catalytic activity of the enzyme (von Meyenburg *et al.*, 1982), and the role of the gene is currently a mystery. The next three genes (*uncB, E, F*) are structural genes for the F_0 and the last five genes are structural genes for the F_1. It is not known how transcription/translation of the genes is regulated such that a final F_1 stoichiometry of $\alpha_3\beta_3\gamma\delta\epsilon$ is achieved. (The same problem exists for F_0 since the uncE protein is present in many more copies than the uncB or uncF proteins). Suggestions involving translational regulation (Brusilow *et al.*, 1982) or selective codon usage (Kanazawa *et al.*, 1981; Gay and Walker, 1981) have been proffered, but are not fully satisfactory. Selective proteolytic degradation of newly translated subunits seems unlikely from data presented (Decker *et al.*, 1982; Klionsky *et al.*, 1983).

Mutations in each of the structural genes have been obtained. From the nature of the selection procedure, these mutations occur at structurally or functionally essential residues in the enzyme. Study of the mutated organisms and of purified enzyme obtained from them gives information regarding both *in vivo* and *in vitro* effects of the mutations. Moreover, from mutant organisms in which the enzyme is totally or almost totally inactivated, partial revertants may be obtained, allowing the investigator to "tailor" the enzyme activity to some extent and potentially giving insight into the interactions of different segments of the polypeptide chains. I shall therefore describe next the contributions that studies of mutants have made to understanding functional mechanism and assembly of the enzyme.

IV. MECHANISM OF PROTON CONDUCTION THROUGH F_0

Dicyclohexylcarbodiimide (DCCD) is a potent inhibitor of proton conduction in F_1F_0-type proton-ATPases, and the work of Beechey showed it to react selectively and covalently with a subunit of F_0 that is hydrophobic ("proteolipid"). This subunit is the uncE protein in *E. coli,* and closely homologous "DCCD-binding proteins" occur in all F_1F_0 proton-ATPases (Sebald and Hoppe, 1981). A diagram of the predicted structure of the *E. coli* uncE protein is shown in Fig. 5 (modified from Senior, 1983). The distribution of secondary structure as predicted in this model is well supported by physical measurements (Mao *et al.*, 1982). Sebald and Hoppe (1981) reviewed the following data which suggested the molecule assumes a "hairpin" arrangement in the membrane. DCCD reacts specifically with residue Asp-61, which is almost certainly buried in the bilayer since DCCD is a hydrophobic molecule, and water-soluble carbodimides do not react with Asp-61 or inhibit proton conduction. A group of "DCCD-resistant" mutants was obtained in which residue 28 (Ile) was mutated to either valine or threonine. These organisms were functionally normal (proton conduction was not altered), but it took larger than usual concentrations of DCCD to inhibit, and the rate of reaction of DCCD with Asp-61 was slowed. It was inferred that Ile-28 may bind DCCD noncovalently prior to the covalent reaction, and the fact that the Ile → Thr mutants were more resistant to DCCD than the Ile-28 → Val mutants was consistent with this concept. Consequently, it was concluded Ile-28 and Asp-61 were closely juxtaposed within the bilayer. The hairpin arrangement is further suggested by the "polarity profile" of the molecule and by secondary structure prediction (Sebald and Hoppe, 1981; Senior, 1983), in that two clearly hydrophobic segments are separated by a central hydrophilic segment which has high β-turn potential. The carboxyl side chain at the position of Asp-61 is conserved in the homologous (proteolipid) subunits of all F_1F_0-type proton-ATPases. Furthermore, *E. coli uncE* mutants in which Asp-61 → Gly, or Asp-61 → Asn, are blocked in proton conduction (Sebald and

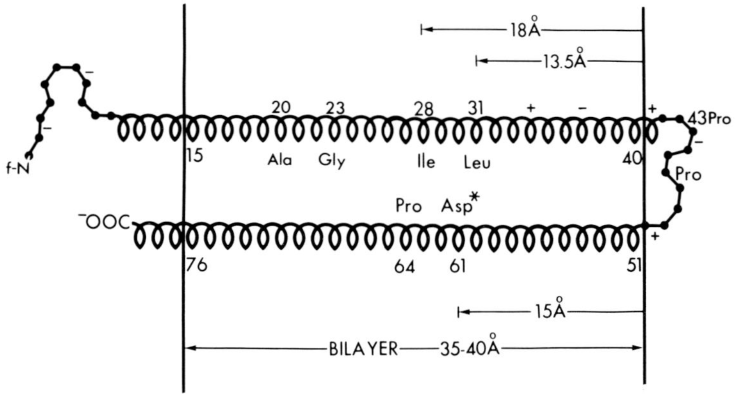

FIG. 5. Secondary and tertiary structure of *E. coli* uncE protein. A random coil or β-turn residue is shown as ●——●, a helical residue is shown by a wavy line. Charged residues are indicated by + or − (Arg, Lys, or carboxyl groups).

Hoppe, 1981; Hoppe *et al.*, 1982), and some mutations in yeast or *N. crassa* which confer resistance to oligomycin (a noncovalent inhibitor of proton conduction in the mitochondrial ATPase) are known to involve amino acid substitutions close to the buried carboxyl group in the proteolipid subunit (Sebald and Hoppe, 1981; see also Senior, 1983). So designation of Asp-61 as a functionally essential residue for proton conduction seems reasonable. From structural considerations of homologous proteins, it may be concluded that Asp-61 is the only conserved, buried, charged residue in the uncE protein.

In contemplating the molecular role of Asp-61, several considerations seem pertinent. First, the carboxyl group may possibly undergo reversible association with H^+. The characteristics of carboxyl protonation–deprotonation within the lipid bilayer milieu are currently obscure. Second, there are multiple copies (6–10) of the uncE protein present per F_0 unit, but it appears DCCD totally blocks proton conduction when only a fraction of the Asp-61 residues (one or two per F_0) have reacted (Sebald and Hoppe, 1981), raising the possibility that not all Asp-61 residues are equivalent in reactivity (deprotonation?), or that some form of cyclical reaction is involved. Third, studies of *E. coli* mutants isolated by Gibson, Cox, and co-workers suggest the environment around the Asp-61 carboxyl group may be of major importance with even relatively small changes causing marked functional aberrations. For example, mutation of Leu-31 → Phe (see Fig. 5 for mutation sites) appears to block proton conduction under some but not all circumstances (Cox *et al.*, 1983). The Phe side chain is nonpolar, but it is considerably larger than Leu, projecting 0.2 nm further from the helix. It is expected to be on the same side of the helix as Ile-28 and may therefore interact with the Asp-61 side chain. Mutation of Pro-64 → Leu (Fimmel *et al.*, 1983) also blocks proton conduction under some conditions. The residue equivalent to Pro-64 is small (Gly, Pro, or Thr) in all the homologous proteins and it seems that substitution of the much bulkier and nonpolar Leu may affect the Asp-61 carboxyl environment, perhaps preventing access of protons from that side. An interesting effect noted by Fimmel *et al.* was that when Pro-64 → Leu, a second-site reversion (Ala-20 → Pro) restored some partial functions.

When reviewing the positions of mutations known to affect proton conduction (residues 61, 31, or 64, see Fig. 5) or known to affect DCCD reactivity (residue 28), one is struck by the way they cluster. Some proposed mechanisms for H^+ conduction through membrane channels envisage a chain of hydrogen bonds providing a pathway across the bilayer. Mutations that block proton conduction might not be expected to cluster together if this were so. However, only a limited number of mutations have been characterized at this time.

To bury a carboxyl group within a lipid bilayer requires energy. In the case of the *E. coli* uncE protein, it appears the nonpolar nature of the majority of the amino acids in the molecule allows Asp-61 to be held in the bilayer. Mutation of Gly-23 → Asp, however, prevents incorporation of the protein into the membrane (Jans *et al.*, 1983), implying that the equilibrium favoring membrane insertion can be made unfavorable by just one extra charged residue per molecule.

Moving briefly now to discuss the role of the other two F_0 subunits (the uncB and uncF proteins) in proton conduction, it is known that both are required to be present for proton conduction to occur (reviewed in Senior and Wise, 1983; see also Loo and Bragg, 1981; Klionsky *et al.*, 1983). Mutations in the uncB protein lead to loss of proton conduction, but in only one case has detailed characterization of the mutation site been reported (Fillingame *et al.*, 1983), and improper assembly of the F_0 seemed a likely consequence of the mutation in that instance. Some speculative inferences regarding a possible essential role of carboxyl residues of the uncB protein in proton conduction have been drawn from considerations of predicted secondary and tertiary structure and from knowledge of loci conferring oligomycin resistance (Senior, 1983), but the kind of detailed evidence upon which firm conclusions can be based is not yet available. Treatment of F_1-depleted normal *E. coli* membranes with proteolytic enzymes gives preferential cleavage of the uncF protein (Hoppe *et al.*, 1983a; Perlin *et al.*, 1983). Proton conduction is not inhibited, indeed it is slightly enhanced (Perlin *et al.*, 1983). These results imply that the membrane-embedded N-terminal segment of the uncF protein (residues 1–30 approximately, see Hoppe *et al.*, 1983a,b; Walker *et al.*, 1982c; Senior, 1983) is sufficient to support proton conduction if properly assembled in F_0. However, no data are available yet to determine which (if any) residues in this segment are of particular functional importance.

V. F_1—THE CATALYTIC UNIT

There is a large body of literature on catalysis in F_1, reviewed by Dunn and Heppel (1981), Amzel and Pedersen (1983), and Senior and Wise (1983). Here I will discuss only experiments that have utilized *E. coli* mutants. The majority of such experiments have not yet gone beyond the stage of defining the F_1 subunit that is altered and describing general characteristics of the defective enzyme.[1] However, detailed studies of *uncA* (α subunit) mutants have provided insight into the catalytic mechanism.

Three *uncA* point mutants (*uncA401, uncA447,* and *unc453*) proved to be of particular interest. It was established that the purified F_1 from each had normal molecular size and subunit composition and that it would re-bind stripped mem-

[1]For work on β-subunit mutants the reader is referred to Senior *et al.* (1979a, 1983), Kanazawa *et al.* (1980), and Wise *et al.* (1983). γ-Subunit mutants have been described by Bragg *et al.* (1973), Downie *et al.* (1980), and Kanazawa *et al.* (1983). δ-Subunit mutants were described by Noumi and Kanazawa (1983) and Humbert *et al.* (1983), and an ε-subunit mutant was described by Cox *et al.* (1977) and Gibson *et al.* (1977).

branes in a normal fashion to block proton conduction through F_0 (Senior *et al.*, 1979b). The enzymes were initially thought to be totally lacking in ATP hydrolytic activity (Wise *et al.*, 1981), although as we realized later, when ATPase assays of high sensitivity are applied, very low levels of activity are detectable.

The α subunit of F_1 carries a nucleotide-binding site that is rather tight and is referred to as "nonexchangeable" or "endogenous." The function of this site is unknown, but it is thought not to be involved in catalysis (Dunn and Heppel, 1981; Perlin *et al.*, 1984). The catalytic site resides on the β subunit (with the possibility that one or more amino acid side chains involved in binding the adenosine moiety of the substrate may be contributed by the α subunit, see Senior and Wise, 1983, for discussion). It was shown that the three noncatalytic (α subunit) and three catalytic (β subunit) nucleotide-binding sites were present in each of the three *uncA* mutant enzymes and, furthermore, that two essential reactive amino acid residues were found to be present with normal reactivity in each of the mutant enzymes (Wise *et al.*, 1981, 1984). An abnormality found in the mutant enzymes was a lack of enhancement of bound aurovertin fluorescence induced by addition of micromolar concentrations of ADP (Wise *et al.*, 1981). Aurovertin is an antibiotic that binds specifically to the β subunit of F_1. On binding it fluoresces, and on addition of ADP, its fluorescence is markedly enhanced in normal F_1. The effect of ADP was lost when normal enzyme was depolymerized into subunits (a process that also inactivated the enzyme), but was regained on repolymerization to an active aggregate. On the basis of these studies we hypothesized (Wise *et al.*, 1981) that an α ↔ β intersubunit conformational interaction was required for normal catalysis in F_1 and that it was blockade of this interaction in the *uncA* mutant F_1 that caused impairment of catalysis. In 1981 we were not yet sure of the location of the site to which ADP actually bound to induce the conformational change. However, it has subsequently become clear that this is the catalytic site on β subunit since both GDP and IDP induce enhancement of bound aurovertin fluorescence while apparently neither binds to the noncatalytic sites on purified F_1 (Perlin *et al.*, 1984).

Figure 6 shows a model to explain these experiments. For simplicity, F_1 is shown as an hexagonal assembly of α and β subunits.[2] On addition of ADP to aurovertin-saturated F_1, a conformational change is transmitted from a catalytic site (where ADP binds) to an aurovertin site (solid and broken line). Only a single ADP binds (mol/mol F_1) under the experimental conditions. (The conformational change may also be measured in reverse as a reduction in F_1–ADP binding affinity caused by binding of aurovertin.) We suggest that the mutations

[2]$\alpha_3\beta_3\gamma$ is the minimal assembly for ATP hydrolysis; only α and β subunits carry nucleotide-binding sites. The γ subunit is currently envisaged to function as an organizer protein for the α and β subunits, whereas δ and ϵ subunits seem to be involved in membrane attachment of F_1 (Dunn and Heppel, 1981).

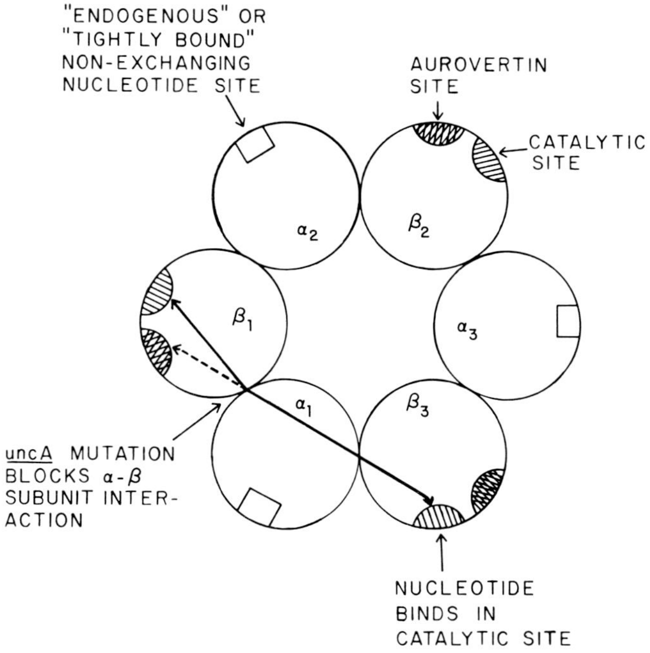

FIG. 6. Site–site interactions in F_1.

in the uncA F_1 enzymes block the $\alpha \leftrightarrow \beta$ intersubunit conformational change, and, as implied in Fig. 6, the mutations may occur at α/β subunit contact faces.

Recent work on mitochondrial F_1 has emphasized the importance of positive catalytic cooperativity in ATP hydrolysis (Grubmeyer and Penefsky, 1981a,b; Grubmeyer *et al.*, 1982; Cross *et al.*, 1982). These workers showed that when only a single catalytic site on F_1 is occupied by substrate, the rate of catalysis is extremely slow, but when two catalytic sites are occupied, the binding at the second site "promotes" the catalysis at the first site, such that a rate enhancement at the first site of up to 10^6-fold occurs. It was apparent, therefore, that the *uncA* mutations might bring about a drastic reduction in catalytic rate by blocking cooperativity between catalytic sites. As Fig. 6 shows (solid line) this idea was consistent with the proposal that $\alpha \leftrightarrow \beta$ intersubunit interaction was blocked in the mutant enzymes. If this were the case, then the *uncA* mutant enzymes should show normal (very low) rates of "single-site catalysis," but the enormous enhancement of catalysis caused when more than one site were occupied by substrate should not be seen.

In normal *E. coli* F_1 very low rates of P_i formation from ATP were seen when only a single site was occupied by substrate, and the measured k for dissociation of P_i under these conditions was close to that reported for mitochondrial F_1 (Wise *et al.*, 1984). When sufficient [^{32}P]ATP was added to saturate only a single catalytic site per *E. coli* F_1 and then allowed to come to equilibrium with bound products, the proportion of [^{32}P]ATP out of the total ^{32}P bound was 0.6 (Wise *et al.*, 1984). This is similar to the value found with mitochondrial F_1 (Grubmeyer *et al.*, 1982). In normal *E. coli* F_1, large enhancement of P_i formation was seen when more than one catalytic site was filled by ATP, confirming that normal *E. coli* F_1 behaves like mitochondrial F_1 in showing marked positive catalytic cooperativity (Wise *et al.*, 1984).

In the *E. coli uncA* mutant F_1 preparations, the rate of P_i formation from ATP when only a single site was occupied was the same as in normal *E. coli* F_1, but on addition of sufficient ATP to saturate more than one site, the rate of P_i formation increased by only small increments in the mutant enzymes (Wise *et al.*, 1984). The positive catalytic cooperativity was therefore greatly attenuated (and to a different extent) in each of the three *uncA* mutant preparations, *uncA401*, *uncA447*, and *uncA453*, confirming that the $\alpha \leftrightarrow \beta$ intersubunit conformational interaction is indeed required for normal catalysis and showing that it mediates the positive catalytic cooperativity, as in Fig. 6.

Partial revertants were obtained from each of these three *uncA* mutants (Senior *et al.*, 1983), and the purified F_1 from the revertants was found to have specific ATPase activity ranging from 1–29% of normal. The $\alpha \leftrightarrow \beta$ intersubunit conformational interaction (measured by the ADP-induced enhancement of bound aurovertin fluorescence) was restored either partially or fully in these revertant enzymes. Thus, with three different mutants and a group of revertant enzymes available for study, it may now be possible to define some of the features of the positive catalytic cooperativity in F_1 at the molecular level.[3]

VI. INTEGRATION OF F_1 AND F_0

How is hydrolysis of ATP coupled to pumping of H^+? How is synthesis of ATP coupled to the energetically downhill progress of H^+ through F_0? Answers to this central question remain elusive. Lacking good hypotheses to test, workers have concentrated more on answering the related, but more elementary, question, how is F_1 bound to F_0?

Evidence summarized by Dunn and Heppel (1981) shows that δ and ϵ subunits of F_1 are required for membrane attachment of F_1. The δ subunit appears to be somewhat rodlike in shape (Sternweis and Smith, 1977) and may have a helical

[3]It should be noted that catalytic site cooperativity in F_1 is an essential feature of ATP synthesis as described by the "binding change mechanism" of P. D. Boyer and co-workers (see Cross, 1981). This mechanism now has considerable experimental support. The three *uncA* mutants described here are defective in ATP synthesis; the revertant organisms described have regained partial capacity for ATP synthesis.

segment about 50 residues long within it (Mabuchi *et al.*, 1981). "OSCP" is a subunit of mitochondrial proton-ATPase, long known to be involved in membrane binding of F_1 (Senior, 1979), and Walker *et al.* (1982b) have recently shown that the sequence of OSCP and *E. coli* δ subunit are homologous. Hoppe *et al.* (1983a) and Perlin *et al.* (1983) showed that the uncF protein of *E. coli* F_0 is involved in F_1 binding, probably through its rather polar C-terminal segment (~120 residues long), which is predicted to be almost entirely helical, and possibly projecting up to 8 nm from the membrane (Walker *et al.*, 1982c; Senior, 1983). Thus we can imagine that in Fig. 2 the uncF protein projects up out of the membrane and that the F_1 subunits are clustered around it. The interaction of the long helical segments of δ subunit of F_1 and uncF protein of F_0 may be important in the functional integration of F_1 and F_0.

VII. ASSEMBLY OF THE *E. coli* PROTON-ATPase

Fully active F_1 may be polymerized *in vitro* from mixtures of individual subunits in the absence of membrane (Dunn and Heppel, 1981). Moreover, the *in vitro* transcription–translation experiments of Downie *et al.* (1981) and Decker *et al.* (1982) showed that the three newly synthesized F_0 subunits would insert spontaneously into *E. coli* membranes or liposomes. From these experiments, from the order of the structural genes in the operon (Fig. 4), and from the *in vitro* reconstitution experiments (Fig. 2), one might infer that assembly of the proton-ATPase *in vivo* occurs via insertion of the F_0 proteins into the membrane to form the proton channel, polymerization of F_1 subunits to an F_1 aggregate in the cytoplasm, and then binding of F_1 to F_0. Other data, however, suggest that this inference would be incorrect.

Cox *et al.* (1981) examined the incorporation of the various subunits of the proton-ATPase into membranes of a series of mutant strains and correlated this with the presence or absence of a functional proton channel in the membranes. Membranes were prepared from whole cells grown in liquid culture. The strains examined were a series of haploid point mutants or polarity mutants, some of them transformed with a multicopy plasmid. The results showed that in order for a functional proton conduction pathway to be formed in membranes, not only must the three F_0 components be present, but the α and β subunits of F_1 must also be membrane bound. The sequence of events in assembly of the enzyme is depicted in Fig. 7. Cox *et al.* (1981) concluded that the α and β subunits regulated insertion of the uncF protein into the membranes *in vivo,* so that the uncF protein was not inserted until after the α and β subunits were bound. With α and β subunits bound, proton conduction through a properly assembled F_0 is blocked, and Cox *et al.* (1981) suggested the assembly sequence was ordered specifically to prevent open F_0 proton channels from debilitating growing cells.

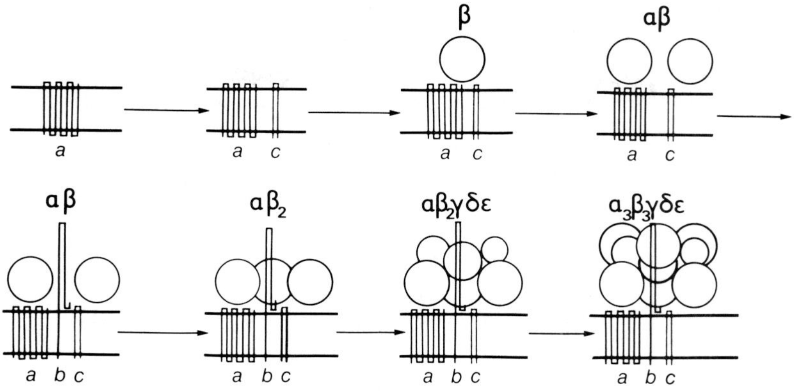

FIG. 7. Assembly of the *E. coli* proton-ATPase *in vivo*. The F_0 subunits are represented by *a* (uncB protein), *b* (uncF protein), and *c* (uncE protein). α, β, γ, δ, and ε are F_1 subunits. The stoichiometry of the F_0 subunits is not known with certainty and the diagram makes no attempt to describe the F_0 morphology. The last three steps should be viewed as quite tentative (Cox *et al.*, 1981). (From Senior and Wise, 1983, with permission.)

Klionsky *et al.* (1983) have studied the assembly of the proton channel in *E. coli* cell and "minicell" membranes using a series of plasmids containing various segments of the *unc* operon to direct transcription–translation. They confirmed that both α and β subunits of F_1 must be synthesized, as well as each of the three F_0 subunits, in order to obtain a functional proton channel, and found that neither δ nor γ subunit of F_1 was required. In some cases, truncated α and β subunits sufficed. However, Klionsky *et al.* disagree with Cox *et al.* on one point, in that they think the uncF protein inserts into the membrane independently of whether α and β subunits of F_1 are present and that the real effect of α and β is to organize the three membrane-inserted F_0 subunits into a functionally competent assembly. The two groups are, however, in agreement on the main point that the proton-ATPase is assembled on the membrane by individual F_1 and F_0 subunits coming together in a concerted program rather than by the binding of preformed F_1 to preformed F_0. Possibly the divergence of opinion is due to effects of gene dosage or proteolysis in the rather different experimental systems used.

VIII. COMPARISON OF THE SUBUNITS OF PROTON-ATPase FROM *E. coli* AND MITOCHONDRIA

Application of genetic and DNA sequencing techniques has revealed homologies between some of the subunits of *E. coli* and mitochondrial proton-ATPases and these are listed in Table I. Also listed in Table I are several subunits that are without known counterparts. Some surprising findings have emerged. For example, the ϵ subunit of mitochondrial F_1 apparently lacks a counterpart in *E. coli*. The uncF protein of *E. coli* has not yet found a counterpart in mitochondrial ATPase (although the subunit called F_6 resembles it vaguely in being highly charged). In general it appears the mitochondrial enzyme has the more complex subunit structure but the reasons for this are currently unclear.

TABLE I
PROTON-ATPase SUBUNITS

E. coli	Mitochondria
Homologous subunits in *E. coli* and mitochondrial enzymes	
α	α[a]
β	β[b]
γ	γ[c]
δ	OSCP[d]
ϵ	δ[d]
uncB protein	ATPase 6[e]
uncE protein	DCCD-binding protein[f]
Subunits without known counterparts	
uncF protein	
	ϵ
	F_6[g]
	F_1-inhibitor protein
	subunit 8[h]
	(Factor B)[i]

[a] Walker *et al.* (1982a).

[b] Runswick and Walker (1983).

[c] J. E. Walker, personal communication.

[d] Walker *et al.* (1982b).

[e] Reviewed in Senior (1983).

[f] Sebald and Hoppe (1981). In mitochondrial F_0 the DCCD-binding protein is also referred to as "the proteolipid subunit" or (in yeast) "subunit 9."

[g] The sequence was kindly communicated personally by E. Racker and R. Bradshaw. Like the 120 residue C-terminal segment of uncF protein, F_6 is highly charged. However, F_6 is only 69 residues long, and little homology to uncF protein was seen here (D. N. Cox, unpublished work).

[h] Macreadie *et al.* (1983). This subunit has been seen only in yeast as yet.

[i] Sanadi (1982). This protein is not definitely proven to be a subunit of the enzyme.

IX. SUMMARY

I have tried in this article to emphasize the ways in which application of genetic techniques has contributed toward understanding the assembly, structure, and function of *E. coli* proton-ATPase, and I think one clear emergent lesson is that the combined application of genetics and biochemical approaches has proved rather useful. It is important to characterize accurately the altered amino acids in the mutants and revertants described in order to maximize the potential information from them, and this, in most cases, remains to be done. Looking ahead, one might consider two important areas of future study to be the determination of the mechanism of functional integration of F_1 with F_0 and the role of the nucleotide site on α subunit. In both cases, site-directed mutagenesis may prove a useful procedure. The differences in structure between the *E. coli* and mitochondrial proton-ATPases are also of interest and at present are of unknown significance. As perception of the wide distribution and varied roles of proton-ATPases in tissues and organisms improves, the impact of the work described here is likely to grow in importance.

ACKNOWLEDGMENTS

I would like to acknowledge the importance of discussions with Professor Frank Gibson and Drs. Graeme B. Cox, John G. Wise, and David S. Perlin in preparing this article, and to acknowledge the excellent experimental work of Lisa Latchney, David Cox, and Anne Ferguson. Work described here from our laboratory in Rochester was supported by NIH Grants GM29805 and GM25349.

REFERENCES

Amzel, L. M., and Pedersen, P. L. (1983). *Annu. Rev. Biochem.* **52,** 801–824.

Apps, D. K. (1982). *Fed. Proc., Fed. Am. Soc. Exp. Biol.* **41,** 2775–2780.

Bragg, P. D., Davies, P. L., and Hou, C. (1973). *Arch. Biochem. Biophys.* **159,** 664–670.

Brusilow, W. S. A., Klionsky, D. J., and Simoni, R. D. (1982). *J. Bacteriol.* **151,** 1363–1371.

Cross, R. L. (1981). *Annu. Rev. Biochem.* **50,** 681–714.

Cross, R. L., Grubmeyer, C., and Penefsky, H. S. (1982). *J. Biol. Chem.* **257,** 12101–12105.

Cox, G. B., Crane, F. L., Downie, J. A., and Radik, J. (1977). *Biochim. Biophys. Acta* **462,** 113–120.

Cox, G. B., Downie, J. A., Langman, L., Senior, A. E., Ash, G. R., Fayle, D. R. H., and Gibson, F. (1981). *J. Bacteriol.* **148,** 30–42.

Cox, G. B., Jans, D. A., Gibson, F., Langman, L., Senior, A. E., and Fimmel, A. L. (1983). *Biochem. J.* **216,** 143–150.

Decker, K. P;, Brusilow, W. S. A., Gunsalus, R. P., and Simoni, R D. (1982). *J. Bacteriol.* **152,** 815–821.

Downie, J. A., Gibson, F., and Cox, G. B. (1979). *Annu. Rev. Biochem.* **48,** 103–131.

Downie, J. A., Langman, L., Cox, G. B., Yanofsky, C., and Gibson, F. (1980). *J. Bacteriol.* **143,** 8–17.

Downie, J. A., Cox, G. B., Langman, L., Ash, G. R., Becker, M., and Gibson, F. (1981). *J. Bacteriol.* **145,** 200–210.

Dunn, S. D., and Heppel, L. A. (1981). *Arch. Biochem. Biophys.* **210,** 421–436.
Fillingame, R. H. (1980). *Annu. Rev. Biochem.* **49,** 1079–1113.
Fillingame, R. H. (1981). *Curr. Top. Bioenerg.* **11,** 35–106.
Fillingame, R. H., Mosher, M. E., Negrin, R. S., and Peters, L. K. (1983). *J. Biol. Chem.* **258,** 604–609.
Fimmel, A. L., Jans, D. A., Langman, L., James, L. B., Ash, G. R., Downie, J. A., Senior, A. E., Gibson, F., and Cox, G. B. (1983). *Biochem. J.* **213,** 451–458.
Gay, N. J., and Walker, J. E. (1981). *Nucleic Acids Res.* **9,** 2187–2194.
Gibson, F. (1983). *Biochem. Soc. Trans.* **11,** 229–240.
Gibson, F., Cox, G. B., Downie, J. A., and Radik, J. (1977). *Biochem. J.* **164,** 193–198.
Gluck, S., Kelly, S., and Al-Awquati, Q. (1982). *J. Biol. Chem.* **257,** 9230–9233.
Goffeau, A., and Slayman, C. W. (1981). *Biochim. Biophys. Acta* **639,** 197–223.
Grubmeyer, C., and Penefsky, H. S. (1981a). *J. Biol. Chem.* **256,** 3718–3727.
Grubmeyer, C., and Penefsky, H. S. (1981b). *J. Biol. Chem.* **256,** 3728–3734.
Grubmeyer, C., Cross, R. L., and Penefsky, H. S. (1982). *J. Biol. Chem.* **257,** 12092–12100.
Haddock, B. A., and Jones, C. W. (1977). *Bacteriol. Rev.* **41,** 47–99.
Harikumar, P., and Reeves, J. P. (1983). *J. Biol. Chem.* **258,** 10403–10410.
Hoppe, J., Schairer, H. U., Friedl, P., and Sebald, W. (1982). *FEBS Lett.* **145,** 21–24.
Hoppe, J., Friedl, P., Schairer, H. U., Sebald, W., von Meyenburg, K., and Jorgensen, B. B. (1983a). *EMBO J.* **2,** 105–110.
Hoppe, J., Montecucco, C., and Friedl, P. (1983b). *J. Biol. Chem.* **258,** 2882–2885.
Humbert, R., Brusilow, W. S. A., Gunsalus, R. P., Klionsky, D. J., and Simoni, R. D. (1983). *J. Bacteriol.* **153,** 416–422.
Jans, D. A., Fimmel, A. L., Langman, L., James, L. B., Downie, J. A., Senior, A. E., Ash, G. R., Gibson, F., and Cox, G. B. (1983). *Biochem. J.* **211,** 717–726.
Kakinuma, Y., Ohsumi, Y., and Anraku, Y. (1981). *J. Biol. Chem.* **256,** 10859–10863.
Kanazawa, H., Horiuchi, Y., Takagi, M., Ishino, Y., and Futai, M. (1980). *J. Biochem.* **88,** 695–703.
Kanazawa, H., Mabuchi, K., Kayano, T., Noumi, T., Sekiya, T., and Futai, M. (1981). *Biochem. Biophys. Res. Commun.* **103,** 613–620.
Kanazawa, H., Noumi, T., Futai, M., and Nitta, T. (1983). *Arch. Biochem. Biophys.* **223,** 521–532.
Klionsky, D. J., Brusilow, W. S. A., and Simoni, R. D. (1983). *J. Biol. Chem.* **258,** 10136–10143.
Loo, T. W., and Bragg, P. D. (1981). *Biochem. Biophys. Res. Commun.* **103,** 52–59.
Mabuchi, K., Kanazawa, H., Kayano, T., and Futai, M. (1981). *Biochem. Biophys. Res. Commun.* **102,** 172–179.
Macreadie, I. G., Novitski, C. E., Maxwell, R. J., John, U., Ooi, B.-G., MacMullen, G. L., Lukins, H. B., Linnane, A. W., and Nagley, P. (1983). *Nucleic Acids Res.* **11,** 4435–4451.
Mao, D., Wachter, E., and Wallace, B. A. (1982). *Biochemistry* **21,** 4960–4968.
Noumi, T., and Kanazawa, H. (1983). *Biochem. Biophys. Res. Commun.* **111,** 143–149.
Perlin, D. S., Cox, D. N., and Senior, A. E. (1983). *J. Biol. Chem.* **258,** 9793–9800.
Perlin, D. S., Latchney, L. R., Wise, J. G., and Senior, A. E. (1984). *Biochemistry* **23,** 4998–5003.
Runswick, M. R., and Walker, J. E. (1983). *J. Biol. Chem.* **258,** 3081–3089.
Sachs, G., Faller, L. D., and Rabon, E. (1982). *J. Membr. Biol.* **64,** 123–135.
Sanadi, D. R. (1982). *Biochim. Biophys. Acta* **683,** 39–56.
Sebald, W., and Hoppe, J. (1981). *Curr. Top. Bioenerg.* **12,** 1–64.
Senior, A. E. (1979). *In* "Membrane Proteins in Energy Transduction" (R. A. Capaldi, ed.), pp. 233–278. Dekker, New York.
Senior, A. E. (1983). *Biochim. Biophys. Acta* **726,** 81–95.
Senior, A. E., and Wise, J. G. (1983). *J. Membr. Biol.* **73,** 105–124.

Senior, A. E., Fayle, D. R. H., Downie, J. A., Gibson, F., and Cox, G. B. (1979a). *Biochem. J.* **180,** 110–118.

Senior, A. E., Downie, J. A., Cox, G. B., Gibson, F., Langman, L., and Fayle, D. R. H. (1979b). *Biochem. J.* **180,** 103–109.

Senior, A. E., Langman, L. P., Cox, G. B., and Gibson, F. (1983). *Biochem. J.* **210,** 395–403.

Senior, A. E., Latchney, L. R., Ferguson, A. M., and Wise, J. G. (1984). *Arch. Biochem. Biophys.* **228,** 49–53.

Shavit, N. (1980). *Annu. Rev. Biochem.* **49,** 111–138.

Simoni, R. D., and Postma, P. W. (1975). *Annu. Rev. Biochem.* **41,** 523–554.

Sternweis, P. C., and Smith, J. B. (1977). *Biochemistry* **16,** 4020–4025.

Stone, D. K., Xie, S.-S., and Racker, E. (1983). *J. Biol. Chem.* **258,** 4059–4062.

Vara, F., and Serrano, R. (1982). *J. Biol. Chem.* **257,** 12826–12830.

von Meyenburg, K., Jorgensen, B. B., Nielsen, J., and Hansen, F. G. (1982). *Mol. Gen. Genet.* **188,** 240–248.

Walker, J. E., Eberle, A., Gay, N. J., Runswick, M. J., and Saraste, M. (1982a). *Biochem. Soc. Trans.* **10,** 203–206.

Walker, J. E., Runswick, M. J., and Saraste, M. (1982b). *FEBS Lett.* **146,** 393–396.

Walker, J. E., Saraste, M., and Gay, N. J. (1982c). *Nature (London)* **298,** 867–869.

Wise, J. G., Latchney, L. R., and Senior, A. E. (1981). *J. Biol. Chem.* **256,** 10383–10389.

Wise, J. G., Duncan, T. M., Latchney, L. R., Cox, D. N., and Senior, A. E. (1983). *Biochem. J.* **215,** 343–350.

Wise, J. G., Latchney, L. R., Ferguson, A. M., and Senior, A. E. (1984). *Biochemistry* **23,** 1426–1432.

Working, P. K., and Meizel, S. (1981). *J. Biol. Chem.* **256,** 4703–4711.

Chapter 9

The Kdp System: A Bacterial K^+ Transport ATPase

WOLFGANG EPSTEIN

Department of Molecular Genetics and Cell Biology
University of Chicago
Chicago, Illinois

I. INTRODUCTION

Studies of *Escherichia coli* have shown that this species has two transport systems capable of accumulating K^+ (17, 32) to achieve the high cellular concentrations—from 0.1 to over 0.6 *M*—found in this species (10). These two systems have some common features but do not appear to have any common components. When *E. coli* is grown in customary media containing excess K^+, only the constitutive Trk system, which has a modest affinity for K^+ (K_m = 1.5 m*M*,) is present. When the Trk system is not able to satisfy their K^+ needs, the

ISBN 0-12-153323-9

cells derepress a second system called Kdp. This system has a very high affinity for K^+ ($K_m = 2\ \mu M$) which allows it to scavenge traces of K^+ in the medium. This article will describe the Kdp system.

The determination of the DNA sequence of the structural genes of the Kdp system has shown it is a member of the class of extensively studied transport ATPases of eukaryotes (3,4,16,20,34). The sequence of KdpB, the largest structural protein of Kdp, is homologous with the Ca^{2+}-ATPase of rabbit sarcoplasmic reticulum (19). To the limited extent that biochemical studies have been done, Kdp appears to have a reaction cycle involving an acyl phosphate intermediate similar to that of the vertebrate transport ATPases. The regulation of Kdp reflects the needs of its bacterial home. Its transport activity is feedback regulated by the turgor pressure, the osmotic pressure difference exerted across the cell membrane. Turgor pressure has also been implicated as the force that is responsible for derepressing the Kdp system when the cells require K^+.

II. STRUCTURE OF THE Kdp SYSTEM

A. Genetic Structure

Kdp was initially identified genetically as the locus of mutations that abolish the ability of the cells to grow in media containing only traces of K^+. Mapping and complementation analysis showed such mutations defined four *kdp* genes, *A* through *D* (11). Recently a fifth gene, *kdpE,* has been identified. These genes are organized into two transcription units or operons as shown in Fig. 1: *kdp* genes *A, B,* and *C* are transcribed as a group from *A* to *C* (31); *kdp* genes *D* and *E* are transcribed from *D* to *E* (28a). The *kdpABC* operon codes for the structural proteins effecting transport, while the *kdpDE* operon codes for two regulatory proteins required to turn on transcription of the *kdpABC* operon to express Kdp transport.

B. Composition of the Kdp Transport Proteins

The *kdpABC* operon has been cloned in a multicopy plasmid and its DNA sequence has been determined (19). The three proteins encoded by this operon form a complex which has been purified with retention of ATPase activity and whose constituent peptides have been isolated after denaturation with SDS (35). The N-terminal sequences of the KdpA and KdpC proteins agree with the sequence predicted from the DNA sequence indicating the proteins do not undergo processing at their N termini (19). Processing at the N terminus is not expected for any of the three structural proteins because none has an N-terminal sequence

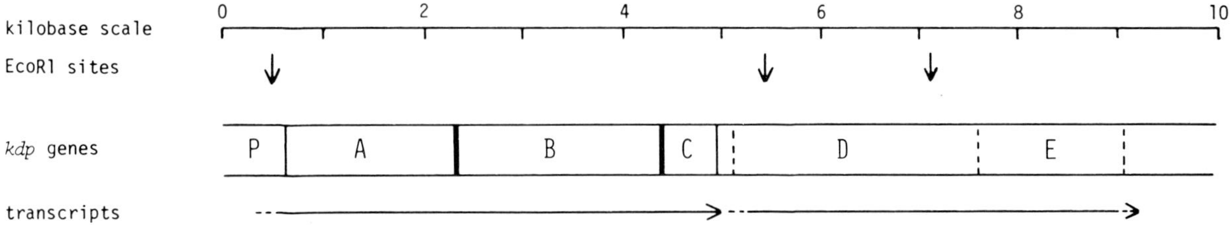

proteins	A	B	C	D	E
amino acid residues	557	682	190	?	?
M_r	59,189	72,112	20,267	95,000	?
location	inner membrane	inner membrane	inner membrane	membrane	?
function	K^+ binding	acyl phosphate intermediate	?	regulates expression	regulates expression

FIG. 1. A diagram of the *kdp* genes of *E. coli*. The line at the top represents the five genes and is drawn to scale for the *kdpA*, *B*, and *C* genes whose size is known. The arrows represent the two transcription units: the *kdpABC* operon codes for the structural genes of the Kdp transport system, and the *kdpDE* operon codes for proteins that regulate expression of the *kdpABC* genes. The sizes of the products of the *A*, *B*, and *C* genes are inferred from the DNA sequence (18), whereas the size of the *kdpD* product is estimated from the electrophoretic mobility of the protein in the presence of SDS.

suggestive of the leader sequences found on most secreted and many membrane-bound proteins.

KdpA is a very hydrophobic 59,189-Da protein (Table I). It contains 66% nonpolar residues and less than 10% charged residues. Hydrophobicity can explain its high electrophoretic mobility in the presence of SDS, corresponding to that of a 47,000-Da protein (23). Hydrophobic proteins are believed to bind disproportionately larger amounts of SDS, thereby migrating more rapidly in an electric field. Hydrophobicity is probably responsible for the rather weak staining of the protein with Coomassie blue and the fact that the band is somewhat more diffuse than those of most proteins. An examination of the sequence shows that charged residues tend to occur in clusters, the intervening portions of 15 to 25 residues being largely hydrophobic (Fig. 2). Since an α helix of about 20

TABLE I
AMINO ACID COMPOSITION OF THE KDP PROTEINS[a]

	KdpA, M_r = 59,189, 557 residues		KdpB, M_r = 72,112, 682 residues		KdpC, M_r = 20,267, 190 residues	
	No.	%	No.	%	No.	%
Neutral, nonpolar	366	65.7	401	58.7	103	54.2
Alanine	62	11.1	97	14.1	24	12.6
Cysteine	6	1.1	6	0.9	0	0.0
Glycine	56	10.0	56	8.2	15	7.9
Isoleucine	27	4.8	52	7.6	10	5.3
Leucine	84	15.1	74	10.9	24	12.6
Methionine	29	5.2	19	2.8	2	1.1
Phenylalanine	35	6.3	24	3.5	5	2.6
Tryptophan	9	1.6	8	1.2	4	2.1
Tyrosine	9	1.6	6	0.9	5	2.6
Valine	49	8.8	59	8.6	14	7.4
Neutral, polar	137	24.6	150	22.0	60	31.6
Asparagine	24	4.3	24	3.5	12	6.3
Glutamine	21	3.8	23	3.4	11	5.8
Proline	25	4.5	24	3.5	13	6.8
Serine	37	6.6	38	5.6	14	7.4
Threonine	30	5.4	41	6.0	10	5.3
Charged, acidic	24	4.3	63	9.3	13	6.9
Aspartic acid	9	1.6	34	5.0	6	3.2
Glutamic acid	15	2.7	29	4.3	7	3.7
Charged, basic	30	5.4	68	9.9	14	7.3
Arginine	16	2.9	32	4.7	8	4.2
Histidine	6	1.1	5	0.7	1	0.5
Lysine	8	1.4	31	4.5	5	2.6

[a] From reference 19.

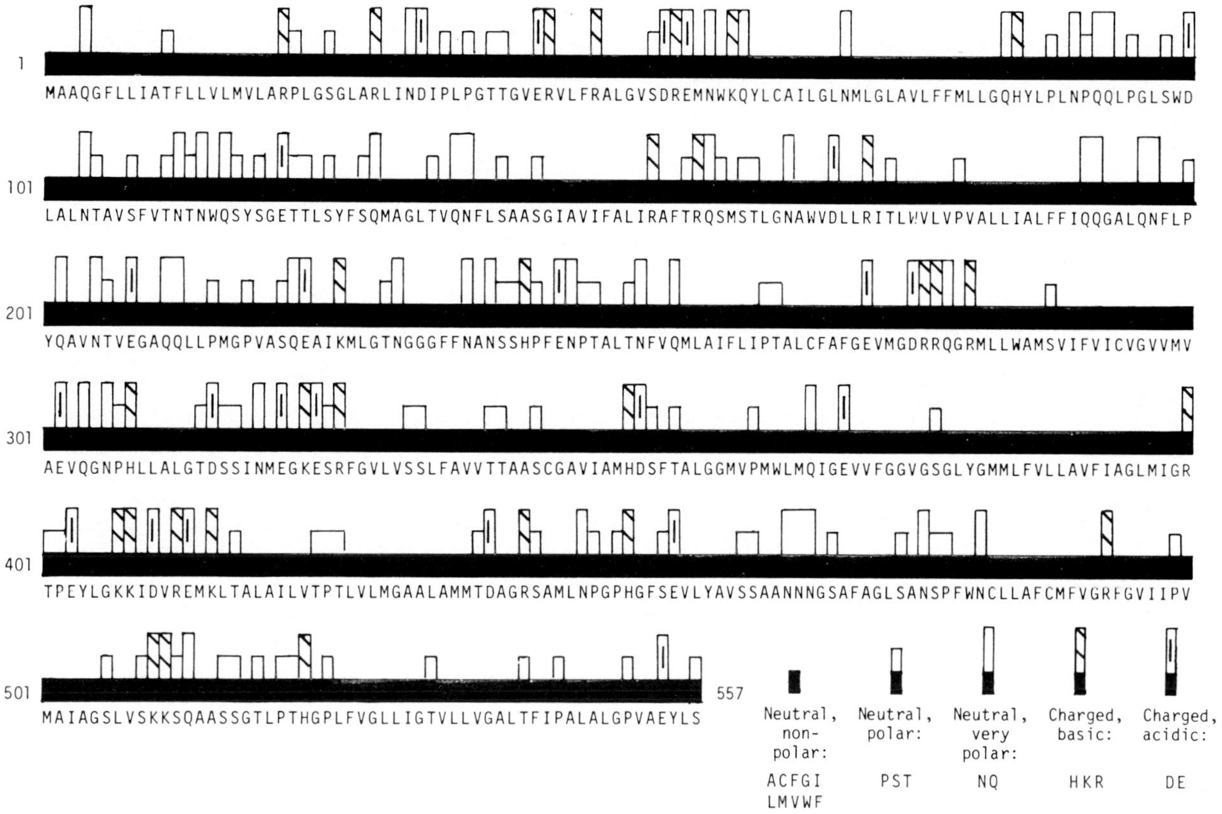

FIG. 2 A diagrammatic presentation of the sequence of the KdpA protein (18). Numbering of residues is from the N terminus beginning at the upper left. The one-letter code for amino acids is used, with symbols as shown at the lower right.

amino acids is long enough to span the lipid region of membranes, many of the hydrophobic stretches in KdpA are candidates for membrane-spanning parts of the protein. The presence of multiple hydrophobic regions suggests KdpA crosses the membrane many times, much like bacteriorhodopsin which has seven membrane-spanning α helices in a protein less than half as large as KdpA (18).

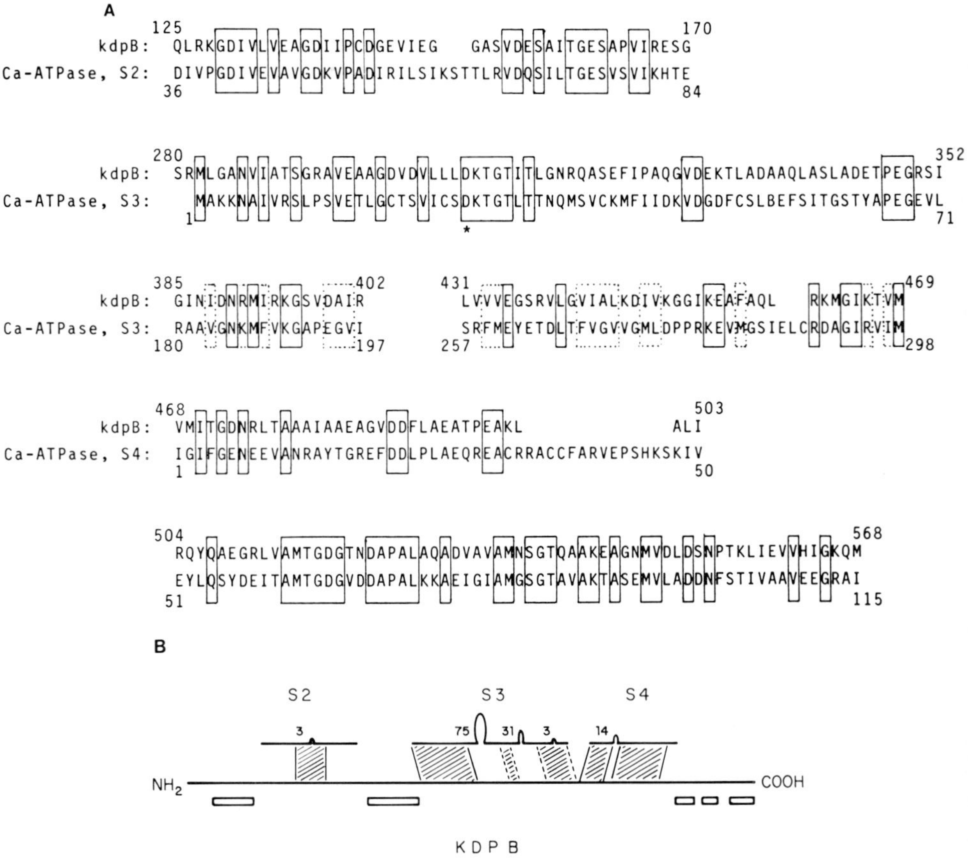

FIG. 3. Sequence homology of KdpB to the Ca^{2+}-ATPase. (A) Homologous regions of the two ATPases. Identical residues are enclosed by a box. Conservative replacements, shown for two less certain homologies to the S3 fragment, are enclosed by dotted lines. The residue indicated by an asterisk, aspartate-26 of the S3 fragment, is the site of phosphorylation of the Ca^{2+}-ATPase (1). (B) The location of the regions of homology in KdpB to three fragments of the Ca^{2+}-ATPase. Regions of homology are joined by shaded bands; those where homology is only suggestive are enclosed by dashed lines. Loops represent places where a deletion of the number of residues shown above the loop will bring the fragment of the Ca^{2+}-ATPase into alignment with KdpB. Boxes represent hydrophobic regions in KdpB which are candidates for membrane-spanning sequences. (From Hesse *et al.*, ref. 19.)

The KdpB protein is the largest of the three with a molecular weight of 72,112 (Table I). It is only moderately hydrophobic. The sequence of this protein has five hydrophobic regions, which are likely to be membrane-spanning (Fig. 3B). The rest of the protein is relatively hydrophilic and would be expected to be in the aqueous compartment outside the membrane. The first two hydrophobic regions, of 50 and 60 residues, respectively, are long enough to span the membrane twice. The three hydrophobic regions near the C terminus are only 20 to 30 residues long; each could span the membrane once. The first of these short residues is somewhat less hydrophobic than the others and so is less likely to be membrane associated. The four or five membrane-spanning regions predicted from the sequence would allow the protein to cross the membranes six or seven times.

The KdpC protein has a molecular weight of 20,267 and is only moderately hydrophobic. Several of its hydrophobic regions (19) could be membrane associated.

C. Topology

The KdpA, B, and C proteins are integral membrane proteins, which can be rendered soluble only by detergents (12,22,23,35). The analysis of amber *kdp* mutants in which the amber fragments appear to be degraded and do not appear in the membrane shows that the absence of any one of these proteins does not prevent the other two from becoming firmly membrane associated (23). Thus at least two of these proteins are integral membrane proteins; the sequences suggest all three are. Dissolution of membranes with nonionic detergents such as Triton X-100 (22) or Aminoxid (35) releases the proteins as a soluble complex. The complex is quite stable; it can be purified by ion-exchange and gel-filtration chromatography with retention of the ATPase activity associated with Kdp.

The stoichiometry of the three subunits in the Kdp transport complex has not yet been established with certainty, but preliminary measurements suggest that each is present in equimolar amounts (22). Genetic evidence indicates that the complex is oligomeric with respect to at least the KdpA protein, containing at least two KdpA proteins. This inference comes from the observation that two *kdpA* missense mutations, *kdpA10* located near the N terminus of the protein and *kdpA4* located near its C terminus (11), complement each other. Such complementation of mutations in the same gene, referred to as intracistronic or interallelic complementation, is considered reliable evidence that the product of the gene is present in an oligomer in which two different malformed subunits can form a functional oligomer (14). This genetic evidence and the probable stoichiometry indicate that the Kdp system has an $A_nB_nC_n$ structure, where n is 2 or more.

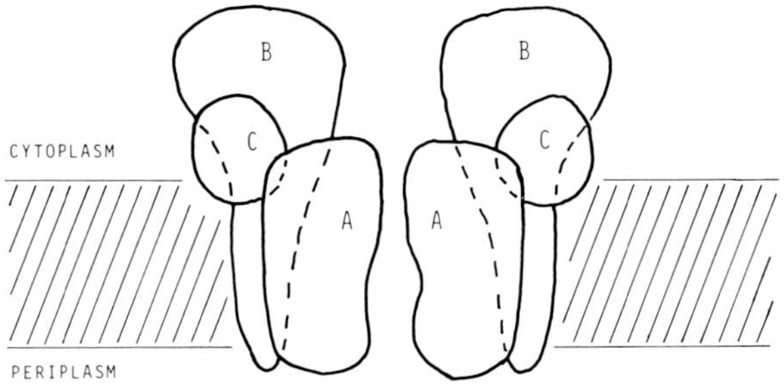

FIG. 4. A schematic representation of the Kdp system, shown as a dimer containing two of each of the three subunits. The evidence for this arrangement is discussed in the text.

Proteases have been used to determine which of the Kdp proteins have extensive regions outside the membrane (22; A. Siebers, personal communication). KdpB is rapidly and extensively degraded by trypsin when inside-out vesicles are used. This result confirms the inference drawn from the primary sequence that a large part of this protein is in the aqueous compartment and indicates that most of this extramembraneous region is in the cytoplasmic compartment. Trypsin can digest KdpB in right-side out vesicles, indicating the protein has trypsin-sensitive sites exposed to the external, periplasmic surface of the membrane. The KdpA and KdpC proteins are refractory to the proteases used to date, in inverted as well as in right-side out vesicles; no evidence on their exposed regions has emerged.

The available information suggests a structure of the Kdp complex shown schematically in Fig. 4. The KdpA protein is shown largely within the membrane. This location is consistent with its resistance to proteases and its hydrophobicity. The complementation between *kdpA* mutations is the basis for drawing two KdpA proteins in contact. Because no intracistronic complementation has been observed among the eight *kdpB* mutations examined, the KdpB subunits may not be in contact with each other. The proteolysis studies are the basis for showing most of KdpB in the cytoplasmic compartment. There is at present no genetic or biochemical evidence for the location of KdpC within the transport complex, so its placement in the figure is arbitrary.

D. Homology with the Ca^{2+}-ATPase

A comparison of the proteins of the Kdp complex with other transport proteins identified only one homologous protein (19), the Ca^{2+}-ATPase of rabbit sarcoplasmic reticulum. The sequences of the homologous regions are shown in Fig. 3A; the locations of homologous regions within KdpB are shown in Fig. 3B. Approximately 65% of the protein sequence of the Ca^{2+}-ATPase has been reported, with some 55% of the sequence assembled into four peptides labeled

S1 to S4, the numbering indicating their order relative to the N terminus (2). The 31-residue N-terminal S1 peptide has no homology, but each of the other three is homologous to a region of KdpB. The regions of homology in KdpB have the same order relative to the N terminus as do the homologous regions of the Ca^{2+}-ATPase peptides. The homologous areas represent parts that must be critical to transport independent of the nature of the transported substrate, which is a monovalent cation for KdpB but a divalent cation for its vertebrate homolog. The homology to the S3 and S4 fragments indicates that part of the size difference between the two proteins is accounted for by multiple small insertions or deletions, depending on whether the common precursor is smaller like KdpB, or larger like the Ca^{2+}-ATPase. Homology of KdpB probably extends to other transport ATPases having an acyl phosphate intermediate, because these are believed to be structurally as well as functionally similar to the Ca^{2+}-ATPase.

III. FUNCTIONS OF Kdp TRANSPORT PROTEINS

A. Kinetics of Kdp Transport

Kdp transport in a wild-type cell is expressed when cells are grown to K^+ limitation (Fig. 5). The curve for cells grown in excess K^+ is that for the constitutive Trk system. In K^+-limited cells an additional component of trans-

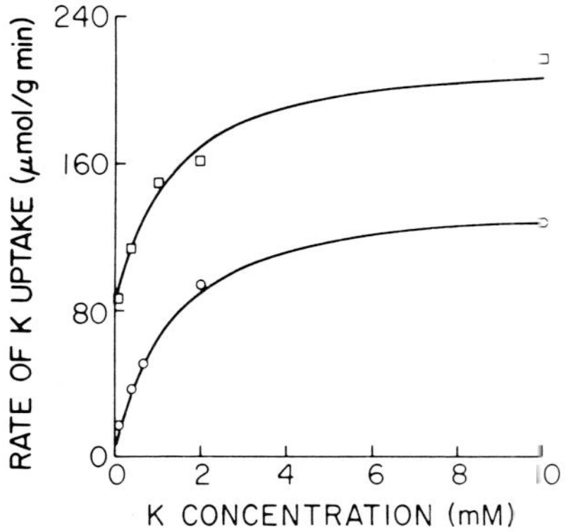

FIG. 5. Kinetics of K^+ transport in a wild-type strain. Strain FRAG-1 was harvested during exponential growth in medium containing 10 m*M* K^+ (○), or 2 hr after depletion of K^+ from medium with an initial K^+ concentration of 0.05 m*M* (□). The initial rate of K^+ uptake was measured in cells suspended in buffer containing chloramphenicol and glucose. Uptake was produced by increasing medium osmolarity with glucose. The lower curve is for a saturable process with a K_m of 1.2 m*M* with a V_{max} of 145 μmol g^{-1} min^{-1}. The upper curve is identical but shifted upward by 80 μmol g^{-1} min^{-1}. (Reproduced from Rhoads *et al.*, *The Journal of General Physiology*, 1976, **67**, pp. 325–341, by copyright permission of The Rockefeller University Press.)

port is seen whose rate is constant over the range 0.1–5 m*M*. This is due to the Kdp system. To study Kdp without confusion with Trk activity, mutants have been isolated that reduce Trk activity to such a low level that Trk activity can be ignored when examining transport or growth below 10 m*M* K^+ (32). In such mutants growth to K^+ limitation fully derepresses Kdp, whereas growth in media containing over 60 m*M* K^+ fully represses Kdp because at these K^+ concentrations the residual Trk activity is adequate for growth.

The affinity of Kdp for K^+ is 2 μ*M*, estimated from the rate of uptake at low concentrations as the cells deplete external K^+ to reach a steady state at an external concentration of about 0.05 μ*M* (32). The high affinity for K^+ is associated with high specificity; Rb^+ and Cs^+ are at best very poor substrates (33), while Tl^+ is transported but with lower affinity than K^+ (8). When the system is fully derepressed it has a rather high maximum rate of transport of 100–150 μmol g^{-1} min^{-1} at 37°C.

B. Energetics of Kdp Transport

The source of energy for Kdp was identified by examining the effect of metabolic inhibitors on activity in intact cells. Arsenate, which interferes with maintenance of high pools of ATP and other high-energy phosphate compounds, inhibited Kdp activity, whereas uncouplers, which reduce the protonomotive force, had relatively little effect (29). The inference, that phosphate bond energy provided energy, was confirmed with the identification of a membrane-bound K^+-stimulated ATPase activity whose properties correlated with those of Kdp transport (12,36). Membranes have a basal ATPase activity that is unrelated to Kdp and indifferent to K^+. In cells in which Kdp activity is depressed, ATPase activity in the presence of K^+ is three to five times the basal level. In two *kdp* mutants in which the affinity of Kdp has been reduced to about 0.3 and 5 m*M*, respectively, the affinity for the ATPase activity is similarly shifted, demonstrating this activity is associated with the Kdp system (Fig. 6).

The cation specificity of a K^+-affinity mutant Kdp-ATPase is shown in Table II. A divalent cation is required; Mg^{2+}, Mn^{2+}, and Co^{2+} are active while Ca^{2+} is inhibitory. When tested at 20 m*M*, the only monovalent cation that stimulates is K^+. Activity is neither inhibited nor stimulated by other ions when present alone or in the presence of K^+.

The ATPase has a reaction cycle in which an unstable phosphoenzyme, an acyl phosphate judging from its stability properties, is formed (13). The intermediate forms rapidly even at 0°C, and turns over rapidly. Gel electrophoresis at low pH showed that the KdpB protein was the site of phosphorylation. Like other ATPases with a phosphorylated intermediate, Kdp-ATPase activity is inhibited by vanadate (28). The radiolabeled intermediate is partially discharged by K^+

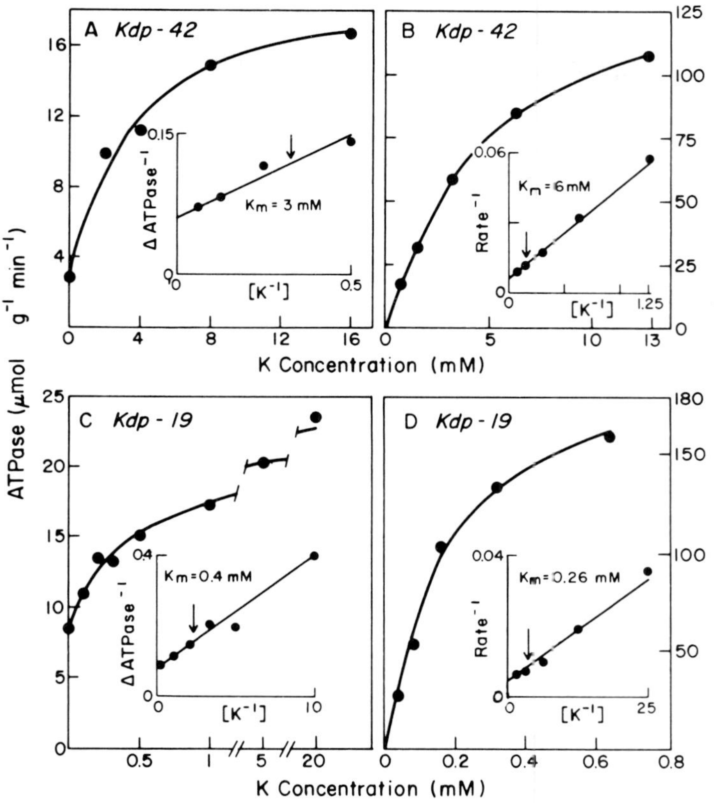

FIG. 6. Correlation of kinetics of the ATPase with kinetics of transport by the Kdp system. Diploid strains carrying the *kdpA42* mutation (A and B) or the *kdpA19* mutation (C and D) were used to measure ATPase in membrane fragments (A and C) and to measure net K^+ uptake (B and D). The inset of each panel presents a double reciprocal plot of the curve. ATPase values are per gram membrane protein; transport rates are per gram of cell dry weight. (Reproduced from Epstein *et al.*, ref. 12.)

(A. Siebers, unpublished observations). In the absence of K^+ it is discharged more rapidly by ADP than by an excess of nonradiolabeled ATP (J. E. Hesse, unpublished observations). This result is analogous to findings for vertebrate transport ATPases, which have an E_1 phosphorylated form in equilibrium with ATP (4,16,20,34). ADP rapidly discharges the E_1 intermediate by reversing its formation, whereas discharge by ATP is slower because it represents turnover of the intermediate through a multistep reaction cycle.

The residue that is phosphorylated in KdpB has not yet been identified. A clue can be obtained from the homology shown in Fig. 3A. One of the regions of homology includes aspartate-26 in fragment S3, the site of phosphorylation in

TABLE II
CATION SPECIFICITY OF THE Kdp-ATPase[a]

A. Divalent cation	Concentration (m*M*)	K^+-stimulated ATPase[b] (μmol g^{-1} min^{-1})
—		0
Mg^{2+}	1.5	23
Mn^{2+}	1.5	21
Co^{2+}	1.5	27
Zn^{2+}	1.5	8
Ca^{2+}	1.5	0
Mg^{2+} and Ca^{2+}	1.5	
	5	2

B. Monovalent cation (20 m*M*)		Total ATPase[c] (μmol g^{-1} min^{-1})
—		9
K^+		54
Li^+		12
Na^+		11
Rb^+		10
Cs^+		10
K^+ and Na^+		53
K^+ and Rb^+		55

[a] Adapted from reference 12.

[b] Activity is the difference between that in the presence of K^+ and in the absence of K^+.

[c] Total membrane ATPase in the presence of 1.5 m*M* Mg^{2+}. Data from a different preparation of membranes with higher activity than for section A.

the Ca^{2+}-ATPase (1). The homologous residue, aspartate-307 of KdpB, is the probable site of phosphorylation.

C. Substrate Binding and Specificity

Mutations that abolish function of the affected gene product, *null* mutations, can identify products needed for a given activity, but do not tell anything about the specific role of the gene product. Mutations producing partial but altered function, *alloiophysic* mutations, allow the identification of a specific function with part of a protein, or with one subunit of a multisubunit complex. To identify the region of the Kdp proteins that determines K^+ binding, mutants having reduced affinity for K^+ were isolated. The selection used, for Kdp activity that allowed growth at a K^+ concentration of 5 m*M* but not at 0.1 m*M*, resulted in the isolation of 17 mutants, of which 11 change only the K_m whereas 6 others change both K_m and V_{max}. The K_m values so obtained range from 0.3 to 100 m*M*, all much higher than the wild-type K_m. Two of the mutants were used to identify the Kdp-ATPase (12). All of the mutations that alter only K_m are located in *kdpA*. Mapping by recombination with small restriction fragments shows the mutations are not clustered but spread out over the gene. Two of the mutations are between residues 100 and 125, another is between residues 470 and 495, and the rest are scattered in between (E. Dorus, unpublished observations). The dispersal of mutations that alter the apparent affinity of the system for K^+ suggests that regions of the protein which are widely separated in the primary structure fold to from the region which determines binding of K^+. The genetic evidence supports inferences from the multiple hydrophobic regions in KdpA that the protein is folded to cross the membrane many times.

In all of the mutants, discrimination against Rb^+ is much reduced; K_m for Rb^+ is from 1 to 30 times that for K^+ (F. Salvacion, unpublished observations). A shift in substrate specificity associated with alterations in K_m supports the notion that this type of mutation alters primarily the site for recognition and binding of the transported cation and does not alter some later step in transport whose kinetic effect is seen as a change in the measured K_m of the process.

IV. REGULATION OF Kdp TRANSPORT

A. Control of Transport by Turgor Pressure

The primary role for the rather high concentrations of K^+ accumulated in *E. coli* appears to be osmoregulation. This species maintains a turgor pressure estimated to be in the range 3–5 atm, corresponding to a difference in osmotic pressure of about 0.2 Osm (21). As shown in Fig. 7, cell K^+ rises with medium osmolarity, suggesting that osmotic balance is maintained by the accumulation of K^+ from the medium. The maintenance of electroneutrality requires the accumulation of anions, or the excretion of cations, equivalent to the cations taken up. Studies of K^+ uptake in response to changes in osmolarity discussed below indicate that K^+ uptake is not associated with excretion of other cations (10). Electroneutrality is maintained by the accumulation of a mixture of metabolically produced acids including glutamate and other organic acids (7,25).

The transport basis for the relationship shown in Fig. 7 is the turgor pressure sensitivity of K^+ transport systems, Kdp and Trk (10,26,30). When turgor pressure is reduced by increasing the osmolarity of the medium, the cells respond by initiating K^+ uptake (10,26,32). This response is immediate, beginning without measurable lag and continuing until the cells have accumulated sufficient K^+ to restore turgor pressure. When the rate of uptake of radiolabeled cation,

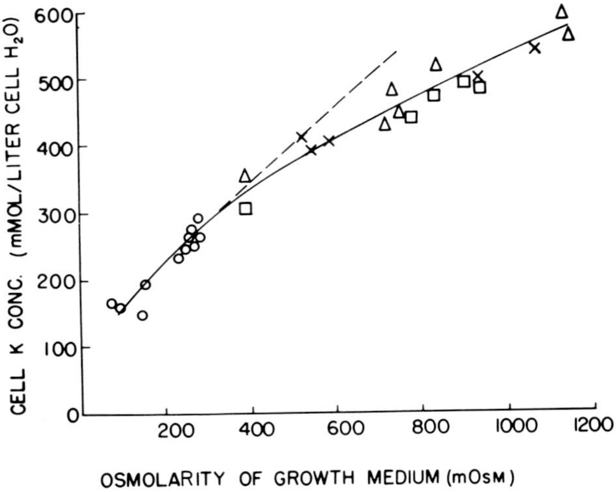

FIG. 7. Dependence of cellular K^+ concentration on osmolarity of the medium. The symbols identify the agent used to adjust osmolarity: ×, glucose; □, NaCl; △, sucrose; ○, no agent added, standard or dilute medium used. The dashed line is the tangent to the curve in the range from 100 to 400 MOsm. (Reproduced from *The Journal of General Physiology*, 1965, **10,** pp. 221–234, by copyright permission of The Rockefeller University Press.)

$^{42}K^+$, is measured simultaneously with net uptake of $^{39}K^+$, it is possible to quantitate unidirectional influx as the rate of uptake of tracer, whereas efflux is calculated from the difference between influx and net flux. This analysis shows that a reduction in turgor pressure stimulates influx but leaves efflux unchanged (Table III). This response to reduced turgor pressure will serve to maintain turgor pressure, accumulating K^+ when turgor pressure is low and ceasing such uptake when optimal levels of turgor pressure are attained. Adjusting internal osmolarity when turgor pressure is too high is mediated by other means, including osmotic

TABLE III

STIMULATION OF K^+ TRANSPORT BY REDUCED TURGOR PRESSURE[a]

System	Condition	Influx[b]	Net flux[b]	Efflux[b]
Kdp	Control	35	0	35
	Reduced Turgor	74	38	36
Trk	Control	23	0	23
	Reduced Turgor	116	92	24

[a] Adapted from reference 30 and reproduced from *The Journal of General Physiology*, 1978, **72,** pp. 283–295, by copyright permission of The Rockefeller University Press.

[b] In μmol g^{-1} min^{-1}.

shock (10). The fact that Trk-mediated transport is also regulated by turgor pressure is consistent with the notion that turgor pressure must be closely regulated and that K^+ transport is the primary mechanism for regulating turgor pressure in *E. coli*.

B. Control of Transport by Ions

The kinetics of transport by Kdp depend on the way transport is measured: whether exchange or net transport is occurring and how net transport is produced. The data of Fig. 8 show progressive inhibition of transport occurs with increasing external K^+ over a range of concentrations well above the K_m for transport. The inhibition is relatively modest when net uptake by K^+-depleted cells is measured. More marked inhibition and a suggestion that V_{max} is lower is noted when uptake in response to a reduction in turgor pressure is measured. The lowest rates of transport and the greatest inhibition by external K^+ occur when exchange is measured. These results show the Kdp system can sense changes in external K^+ concentrations orders of magnitude above the K_m and discriminate against the high concentrations of Na^+ present in all of the experiments of Fig. 8. The response is modulated by turgor pressure and internal ionic content, with maximum inhibition when turgor pressure is high and ionic content is that typical

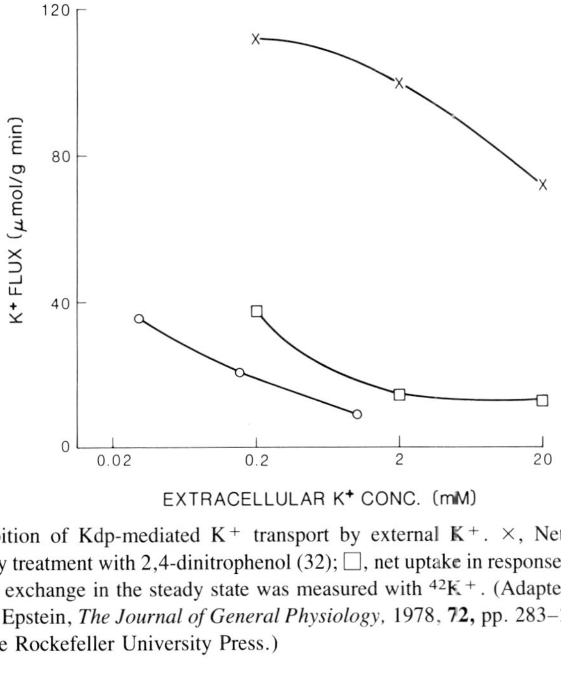

FIG. 8. Inhibition of Kdp-mediated K^+ transport by external K^+. ×, Net uptake by cells depleted of K^+ by treatment with 2,4-dinitrophenol (32); □, net uptake in response to reduced turgor pressure; ○, K^+ exchange in the steady state was measured with $^{42}K^+$. (Adapted and reproduced from Rhoads and Epstein, *The Journal of General Physiology*, 1978, **72**, pp. 283–295, by copyright permission of The Rockefeller University Press.)

of growth. The maximum rate of uptake in response to a reduction of turgor pressure may be lower than that seen in K^+-depleted cells because in the former case the cells must synthesize the charge-balancing anions, whereas in the latter most of the uptake is an exchange for internal Na^+.

C. Regulation of Kdp Expression by Turgor Pressure

In a wild-type cell Kdp is expressed only when cells are grown in media containing low concentrations of K^+ (Fig. 5). However, when expression was studied in mutants in which Trk activity was very low, partial expression of Kdp occurred during growth in media containing as much as 10 m*M* K^+ (32). To examine regulation in more detail a transcriptional fusion between the *kdpABC* operon and the gene for β-galactosidase was isolated (24). This fusion was obtained by insertion of the Mu(*Ap, lac*) phage (5) into the *kdpA* gene. The insertion abolishes expression of Kdp transport while putting expression of β-galactosidase under control of the *kdpABC* promoter. In this way transcriptional regulation of the *kdpABC* operon can be measured in the absence of Kdp function.

The expression of Kdp in this strain is dependent on the activity of Trk, as inferred earlier (Fig. 9). In strain TL1110 with wild-type Trk activity, expression is not seen above 5 m*M* K^+ and achieves maximal rates only below 1 m*M*. By contrast strain TL1105 with markedly reduced Trk activity expresses the system slightly at 50 m*M* K^+ and maximally below 20 m*M*. These results show that external K^+ does not regulate expression directly. The advantage of studying expression in the absence of Kdp function is illustrated by the curves with open symbols in Fig. 9, which represent the same strains but with a wild-type Kdp system introduced on an F-episome. Expression begins over the same range of K^+ concentrations as in the strain lacking Kdp activity, but maximum expression is only 10% of that in the absence of Kdp function. Thus the level of expression is markedly reduced when Kdp function is intact.

To examine the role of internal K^+ concentration, we made use of the osmotic dependence of this parameter (Fig. 7). Experiments like those of Fig. 9 were performed with strain TL1105 growing in media of lower osmolarity, so that cell K^+ was lower (24). In each medium it was found that there was an external K^+ concentration below which expression was progressively turned on, but the concentration at which this occurred varied with osmolarity. Whether internal K^+ was only 130 m*M* (in 45 mOsm medium), 160 m*M* (in 99 mOsm medium) or 225 m*M* (in 198 MOsm medium), expression was still determined by external K^+. An examination of growth rates in these experiments showed that expression began just above the external K^+ concentration at which growth began

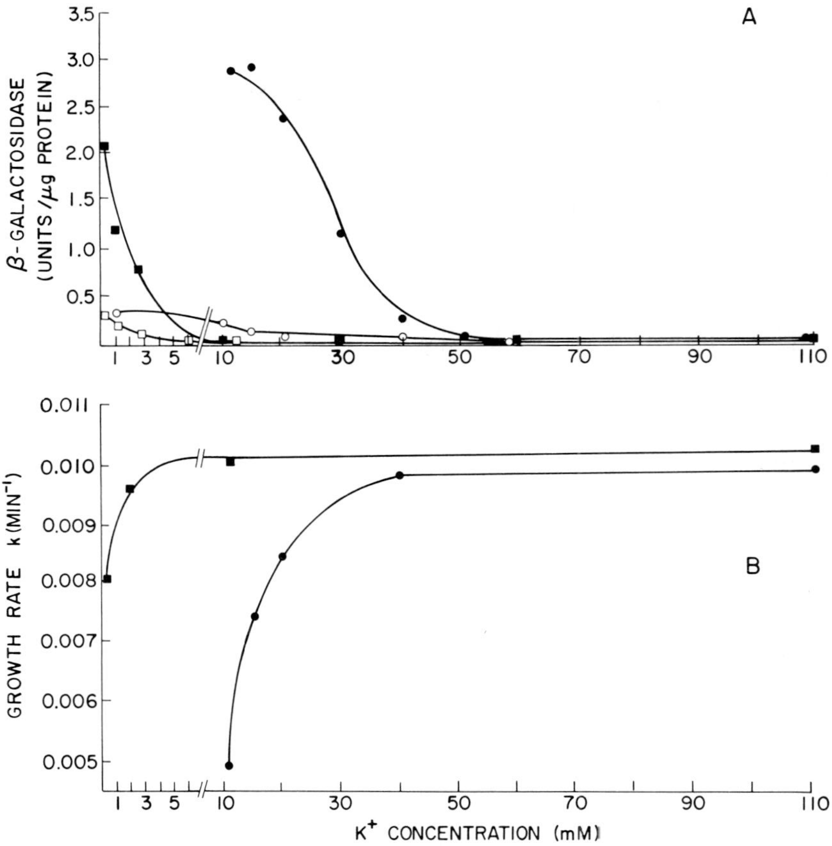

FIG. 9. Effect of external K^+ concentration on Kdp expression and growth rate in strains with a transcriptional *kdp::lacZ* fusion. (A) Steady-state expression of β-galactosidase. (B) Growth rates in the same experiments. Strains lacking a functional Kdp system: ●, TL1105 (Trk^-); ■, TL1110 (Trk^+). Strains with a functional Kdp system in trans on the F-100 episome: ○, TL1106A (Trk^-); □, TL111A (Trk^+). (From Laimins *et al.*, ref. 24.)

to fall, the point at which the rate of uptake of K^+ became rate limiting for growth. The parameter the cells seem to sense is their need for K^+, not its absolute concentration inside or outside.

The osmotic role of K^+ suggested that turgor pressure might be the way the need for K^+ was sensed to regulate Kdp expression. This possibility was tested by measuring expression in response to a reduction in turgor. As shown in Fig. 10A, there was a burst of β-galactosidase synthesis after turgor pressure was reduced under conditions where no expression occurred when turgor pressure was normal. Because reducing turgor pressure also has transient effects on protein synthesis, the incorporation of an amino acid was measured at the same time (Fig. 10B). The plot of expression versus protein synthesis (Fig. 10C) shows a short lag, then a period of expression at a constant rate, and later gradual turning off of expression. Under these conditions there has been no change in external K^+, and initially cell K^+ either remained constant or increased if the change in osmolarity was large enough to produce plasmolysis with osmotic exit of water from the cell. Transient expression of Kdp, produced with glucose in Fig. 10, was observed with all osmotically active solutes tested, but not with glycerol which is known to permeate rapidly (15).

These data fit well with a model in which turgor pressure is the determinant of Kdp expression. When turgor pressure is at optimal levels, there is no expression of Kdp. When turgor pressure falls below these levels, there is progressive derepression of Kdp. A slight reduction in turgor does not have any other apparent effect, but larger reductions reduce growth rate. In cells without a functional Kdp system, modest expression of Kdp occurs at intermediate K^+ concentrations where uptake is not able to maintain the optimal turgor pressure but where the reduction does not affect growth rate. At lower K^+ concentrations turgor pressure falls to levels that reduce growth rate and Kdp expression rises to high levels. In cells with a functional Kdp system turgor pressure falls only enough to produce modest expression of Kdp. The level of expression allows a rate of K^+ transport that maintains turgor pressure high enough to allow for a normal growth rate but slightly below the optimum so as to maintain partial expression of Kdp.

Genetic studies implicate the products of the *kdpD* and *kdpE* genes in the regulation of Kdp. Expression of the *kdpABC* operon requires that the cell have functional *kdpD* and *kdpE* genes (23, 28a). Insertion and nonsense mutations in *kdpD* and many point mutations in *kdpD* and *kdpE* block expression of the Kdp system. All of these mutations are recessive to the wild-type. Mutations to partial constitutive expression have been isolated; these are dominant to the wild-type and map in or near the *kdpDE* genes (31). The genetic data indicate that *kdpD* and *kdpE* code for positive regulators of Kdp expressions. Studies of plasmids and transducing bacteriophages carrying all of *kdpD* and part of *kdpE* indicate that *kdpD* codes for a 95,000-Da membrane-associated protein (D. Dosch and J. E. Hesse, unpublished observations). The entire *kdpDE* operon has now been cloned in a bacteriophage as well as a multicopy plasmid.

Regulation of Kdp expression would seem to require two separate events: the sensing of turgor pressure and the conversion of this information to effect initiation of transcription at the *kdpABC* promoter. The simplest model would have these functions performed by the KdpD and KdpE proteins alone, without other molecules. The membrane location of KdpD is consistent with a component that senses turgor pressure. A membrane protein could alter its conformation in response to the pressure exerted across the cell envelope and/or to stretch in the

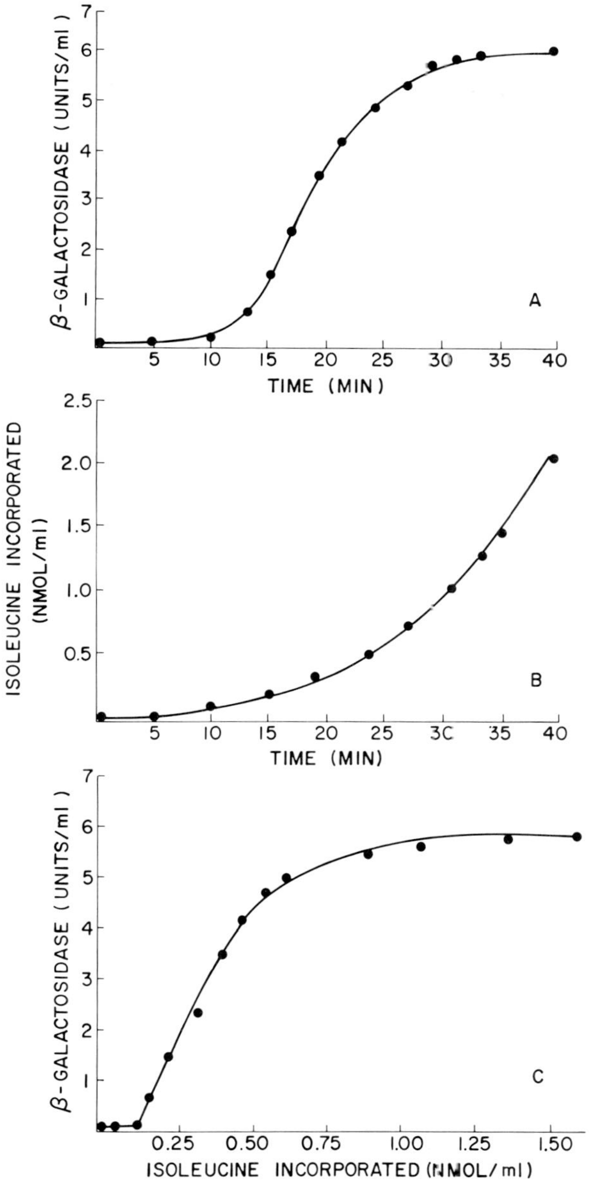

FIG. 10. Effect of shift in medium osmolarity on *Kdp* expression in strain TL1105 (described in the legend to Fig. 9). At time zero, cells growing in medium containing 60 m*M* K^+ received radiolabeled isoleucine and 0.11 vol 2.1 *M* glucose in medium. (A) Expression of β-galactosidase as a function of time. (B) Incorporation of isoleucine as a function of time. (C) Differential plot of β-galactosidase synthesis versus isoleucine incorporation. (From Laimins *et al.*, ref. 24.)

plane of the envelope produced by turgor pressure. The failure to identify other genes that alter regulation of Kdp does not exclude them. The localization of the two identified regulatory proteins and biochemical studies of their function are a necessary first step in unraveling this unusual control mechanism.

V. QUESTIONS AND CHALLENGES

Kdp presents challenges in two rather different areas: the mechanism of ATP-driven transport and the mechanism underlying regulation by turgor pressure. This overview suggests some of the biochemical and genetic studies needed to support models based on preliminary results. The biochemical information needed includes the stoichiometry of the complex and a complete characterization of the nature, formation, and discharge of the phosphorylated intermediate. Purification of the complex to homogeneity will allow reconstitution in vesicles to determine if transport is coupled to movement of another ion and the stoichiometry of transport. Further genetic analysis may help identify specific functions of the subunits and of regions of the subunits.

Some conclusions can be drawn which may be applicable to, or contrasted with, findings for other transport ATPases. Intracistronic complementation among mutations affecting the KdpA subunit shows the functional complex is oligomeric for this subunit. Considerable data suggest that several other ATPases are oligomeric (4,20,34), but monomeric forms of detergent-solubilized enzymes retain ATPase activity similar to that of the native enzyme (6,27). Sites for cation binding and translocation must be on the same subunit that binds ATP in the case of those ATPases that have only a single subunit. In Kdp, genetic evidence implicates the KdpA subunit in K^+ binding, whereas a different subunit, KdpB, is the site of phosphorylation. The structure of KdpA suggests it is multiply folded across the membrane, an arrangement that could well form a transmembrane path for cation movement. If so, KdpB (and KdpC) would serve to couple energy by a conformational change produced by ATP. Does the small β subunit of the Na^+,K^+-ATPase serve a role similar to that of KdpA? This is suggested by analogy with Kdp, but the β subunit is quite hydrophilic (4) and so is unlikely to have a structure or function like that of KdpA.

Unique to Kdp is regulation by turgor pressure. The effect on transport (Table III, refs. 10,26,30), as distinguished from control of gene expression, must be exerted on the proteins of the transport system. This response is the analog of feedback inhibition of key steps in catabolic and anabolic pathways. Here the "product" is turgor pressure, controlling the key step in changing turgor. Because both Kdp and Trk systems exhibit this type of control, it is possible that such control is mediated by a single sensor that interacts with both systems. No such common components have been identified, but the selections used may well not have enriched for this sort of a mutant. It is also possible that this response is inherent in each transport system, so that the sensor of turgor pressure for Kdp activity is in one of its three subunits.

Equally unusual is the way Kdp expression is controlled. Osmotic pressure has been shown to regulate synthesis of porins and other outer membrane proteins, and the synthesis of an anionic polysaccharide in *E. coli,* but these phenomena represent control by osmotic pressure of the medium (9), not by turgor pressure, which is the difference between medium and cellular osmolarity. The evidence for regulation by turgor pressure is circumstantial; such control fits the data well. Genetics has identified two regulatory proteins, KdpD and KdpE, which are required. One of these, KdpD, is membrane associated. The site where these proteins act, the *kdpABC* promoter, has not been sequenced; the region that has been sequenced includes only part of the promoter (19). The characterization of the proteins, of the promoter, and of interactions of proteins with the promoter are needed before concrete models for such regulation can be taken seriously. With luck, it may be possible to reconstitute regulation *in vitro.*

ACKNOWLEDGMENTS

The author thanks his colleagues in Chicago: E. Dorus, D. Dosch, J. E. Hesse, and J. W. Polarek, and his colleagues in Osnabrück: K. Altendorf, A. Siebers, and L. Wieczorek, for providing unpublished data cited in this article. Work on the Kdp system in the author's laboratory was supported by Grant GM 22323 from the National Institutes of Health and Grants PCM 75-14016 and 7904641 from the National Science Foundation. This review was begun during a pleasant sabbatical visit to Osnabrück supported in part by a fellowship from the Alexander von Humboldt Foundation.

REFERENCES

1. Allen, G., and Green, N. M. (1976). A 31-residue tryptic peptide from the active site of the [CA^{++}]-transporting adenosine triphosphatase of rabbit sarcoplasmic reticulum. *FEBS Lett.* **63,** 188–192.
2. Allen, G., Trinnaman, B. J., and Green, N. M. (1980). The primary structure of the calcium ion-transporting adenosine triphosphatase protein of rabbit skeletal sarcoplasmic reticulum. *Biochem. J.* **187,** 591–616.
3. Briskin, D. P., and Leonard, R. T. (1982). Partial characterization of a phosphorylated intermediate associated with the plasma membrane ATPase of corn roots. *Proc. Natl. Acad. Sci. U.S.A.* **79,** 6922–6926.
4. Cantley, L. C. (1981). Structure and mechanism of the (Na,K)-ATPase. *Curr. Top. Bioenerg.* **11,** 201–237.
5. Casadaban, M. J., and Cohen, S. N. (1979). Lactose genes fused to exogenous promoters in one step using a Mu-*lac* bacteriophage: *In vivo* probe for transcriptional control sequences. *Proc. Natl. Acad. Sci. U.S.A.* **76,** 4530–4533.
6. Craig, W. S. (1982). Monomer of sodium and potassium activated adenosinetriphosphatase displays complete enzymatic function. *Biochemistry* **21,** 5707–5717.
7. Damadian, R. (1971). Biological ion exchange resins. I. Quantitative electrostatic correspondence of fixed charges and mobile counter ions. *Biophys. J.* **11,** 739–760.

8. Damper, P. D., Epstein, W., Rosen, B. P., and Sorensen, E. K. (1979). Thallous ion is accumulated by potassium transport systems in *Escherichia coli. Biochemistry* **18,** 4165–4169.
9. Epstein, W. (1983). Membrane-mediated regulation of gene expression in bacteria. *In* "Gene Function in Prokaryotes" (J. Beckwith, J. Davies, and J. A. Gallant, eds), pp. 281–292. Cold Spring Harbor Laboratory, Cold Spring Harbor, New York.
10. Epstein, W., and Schultz, S. G. (1965). Cation transport in *Escherichia coli* V. Regulation of cation content. *J. Gen. Physiol.* **49,** 221–234.
11. Epstein, W., and Davies, M. (1970). Potassium-dependent mutants of *Escherichia coli. J. Bacteriol.* **101,** 836–843.
12. Epstein, W., Whitelaw, V., and Hesse, J. (1978). A K^+ transport ATPase in *Escherichia coli. J. Biol. Chem.* **253,** 6666–6668.
13. Epstein, W., Laimins, L. A., and Hesse, J. E. (1979). A phosphorylated intermediate of the kdp system, an ATP-driven K^+ transport system of *E. coli. Abstr. Int. Congr. Biochem. 11th Toronto* p. 449.
14. Fincham, J. R. S. (1959). On the nature of the glutamate dehydrogenase produced by inter-allele complementation at the *am* locus of *Neurospora crassa. J. Gen. Microbiol.* **21,** 600–611.
15. Fischer, A. (1903). "Vorlesungen über Bacterien," 2nd Ed., p. 25. Fischer, Jena.
16. Goffeau, A., and Slayman, C. W. (1981). The proton-translocating ATPase of the fungal plasma membrane. *Biochim. Biophys. Acta* **639,** 197–223.
17. Helmer, G. L., Laimins, L. A., and Epstein, W. (1982). Mechanisms of potassium transport in bacteria. *In* "Membranes and Transport" (A. N. Martonosi, ed.), Vol. 2, pp. 123–128. Plenum, New York.
18. Henderson, R., and Unwin, P. N. T. (1975). Three-dimensional models of purple membrane obtained by electron microscopy. *Nature (London)* **257,** 28–32.
19. Hesse, J. E., Wieczorek, L., Altendorf, K., Reicin, A. S., Dorus, E., and Epstein, W. (1984). Sequence homology between two membrane transport ATPases, the Kdp-ATPase of *E. coli* and the Ca^{2+}-ATPase of sarcoplasmic reticulum. *Proc. Natl. Acad. Sci. U.S.A.* **81,** 4746–4750.
20. Ikemoto, N. (1982). Structure and function of the calcium pump protein of sarcoplasmic reticulum. *Annu. Rev. Physiol.* **44,** 297–317.
21. Knaysi, G. (1951). "Elements of Bacterial Cytology," 2nd Ed., p. 155. Comstock, Ithaca, New York.
22. Laimins, L. A. (1981). Organization of the KDP potassium transport system of *Escherichia coli.* Ph.D. thesis, University of Chicago.
23. Laimins, L., Rhoads, D. B., Altendorf, K., and Epstein, W. (1978). Identification of the structural proteins of an ATP-driven potassium transport system in *Escherichia coli. Proc. Natl. Acad. Sci. U.S.A.* **75,** 3216–3219.
24. Laimins, L. A., Rhoads, D. B., and Epstein, W. (1981). Osmotic control of *kdp* operon expression in *Escherichia coli. Proc. Natl. Acad. Sci. U.S.A.* **78,** 464–468.
25. Measures, J. C. (1975). Role of amino acids in osmoregulation of non-halophilic bacteria. *Nature (London)* **257,** 398–400.
26. Meury, J., and Kepes, A. (1981). The regulation of potassium fluxes for the adjustment and maintenance of potassium levels in *Escherichia coli. Eur. J. Biochem.* **119,** 165–170.
27. Møller, J. V., Lind, K. E., and Andersen, J. P. (1980). Enzyme kinetics and substrate stabilization of detergent-solubilized and membraneous (Ca^{2+} + Mg^{2+})-activated ATPase from sarcoplasmic reticulum. *J. Biol. Chem.* **255,** 1912–1920.
28. O'Neal, S. C., Rhoads, D. B., and Racker, E. (1979). Vanadate inhibition of sarcoplasmic reticulum Ca^{2+}-ATPase and other ATPases. *Biochem. Biophys. Res. Commun.* **89,** 845–850.
28a. Polarek, J. W., Williams, G., Hesse, J. E. and Epstein, W. (1985). In preparation.
29. Rhoads, D. B., and Epstein, W. (1977). Energy coupling to net K transport in *Escherichia coli* K-12. *J. Biol. Chem.* **252,** 1394–1401.

30. Rhoads, D. B., and Epstein, W. (1978). Cation transport in *Escherichia coli.* IX. Regulation of K Transport. *J. Gen. Physiol.* **72,** 283–295.
31. Rhoads, D. B., Laimins, L. A., and Epstein, W. (1978). Functional organization of the *kdp* genes of *Escherichia coli. J. Bacteriol.* **135,** 445–452.
32. Rhoads, D. B., Waters, F. W., and Epstein, W. (1976). Cation transport in *Escherichia coli.* VIII. Potassium transport mutants. *J. Gen. Physiol.* **67,** 325–341.
33. Rhoads, D. B., Woo, A., and Epstein, W. (1977). Discrimination between Rb and K by *Escherichia coli. Biochim. Biophys. Acta* **469,** 45–51.
34. Sachs, G., Wallmark, B., Saccomani, G., Rabon, E., Stewart, H. B., DiBona, D. R., and Berglindh, T. (1982). The ATP-dependent component of gastric acid secretion. *Curr. Top. Membr. Transp.* **16,** 135–159.
35. Wieczorek, L. (1983). Thesis, Universität Bochum.
36. Wieczorek, L., and Altendorf, K. (1979). Potassium transport in *Escherichia coli.* Evidence for a K^+-transport adenosine-5′-triphosphatase. *FEBS Lett.* **98,** 233–236.

Chapter 10

Molecular Cloning and Characterization of a Mouse Ouabain Resistance Gene: A Genetic Approach to the Analysis of the Na^+,K^+-ATPase

ROBERT LEVENSON[1]

Center for Cancer Research and
Department of Biology
Massachusetts Institute of Technology
Cambridge, Massachusetts

I. INTRODUCTION

Cell proteins that carry out ion transport are an important class of proteins that have been particularly difficult to study by conventional protein chemistry techniques. From a general perspective, the application of somatic cell genetics and

[1]Present address: Department of Cell Biology, Yale University School of Medicine, New Haven, Connecticut 06510.

ISBN 0-12-153323-9

recombinant DNA technology to this class of proteins has the potential to open new avenues of approach to the study of these proteins. Purification of the polypeptide chain responsible for transport of a particular ion continues to be a technically difficult goal to achieve, especially in situations where such proteins are present in low abundance. In cases where drug-resistant mutants can be selected, DNA transfer techniques could provide an alternate route to isolation of the DNA sequences coding for such proteins. This article will focus on the use of DNA-mediated gene transfer as an experimental approach for isolating the DNA sequence coding for the Na^+,K^+-ATPase subunit responsible for ouabain resistance.

The plasma membrane Na^+,K^+-ATPase is the enzymatic activity responsible for the maintenance of the high internal K^+ and low internal Na^+ concentrations characteristic of most living cells. Although the Na^+,K^+-ATPase has been purified, an understanding of the detailed molecular structure of the enzyme remains a challenging biochemical problem. The ability to isolate the gene coding for the ATPase would represent an important technical step toward elucidating the primary structure of the Na^+,K^+-ATPase.

Our approach for isolating the Na^+,K^+-ATPase gene is based on three main considerations: (1) the cardiac glycoside ouabain is a specific inhibitor of the Na^+,K^+-ATPase (Cantley, 1981), and cell lines of different species vary considerably with respect to ouabain sensitivity (Baker, 1976); (2) ouabain-resistant mutants exhibit decreased affinity for ouabain (Robbins and Baker, 1977); (3) ouabain resistance can be transferred from ouabain-resistant to ouabain-sensitive cells by direct DNA transfer (Corsaro and Pearson, 1981), metaphase chromosome transfer (Baker and Gross, unpublished), or by microcell fusion (Kozak *et al.*, 1979).

We have taken advantage of the species-specific differences in ouabain sensitivity to develop a gene transfer and selection system for isolating the mouse gene responsible for ouabain resistance. Since biochemical experiments indicate that the α subunit of the ATPase binds ouabain (Ruoho and Kyte, 1974), we presume that the DNA sequence that confers ouabain resistance codes for the α subunit of the Na^+,K^+-ATPase. The purpose of this article is to outline the methodologies used to isolate the ouabain resistance gene and to describe how this DNA sequence can be used to facilitate genetic studies of the Na^+,K^+-ATPase and the ouabain resistance locus.

II. STRATEGY FOR ISOLATION OF THE OUABAIN RESISTANCE GENE

A. DNA-Mediated Gene Transfer and Construction of Ouabain-Resistant Transformed Cell Lines

Experiments carried out by Baker and co-workers (Baker, 1976) demonstrated that cell lines of different species vary considerably with respect to the level of ouabain that is toxic to wild-type cells. In particular, cells of human origin were shown to be quite sensitive to ouabain compared to rodent cells (Baker, 1976). We have found that primate cells exhibit sensitivity to ouabain at levels comparable to human cells (Levenson, *et al.*, 1984). A dose of 5×10^{-8} *M* ouabain is sufficient to cause greater than 95% loss of viability when applied to African green monkey CV-1 cells, whereas mouse cells are resistant to a dose of 2×10^{-4} *M* ouabain. This species-specific variation in ouabain sensitivity has made it possible to develop a gene transfer and selection system for isolating the mouse ouabain resistance gene.

The experimental scheme involved in this procedure is outlined in Fig. 1. The rationale behind this approach is based on the idea that DNA from ouabain-resistant cells can transform ouabain-sensitive cells to ouabain resistance employing DNA-mediated gene transfer techniques. In this procedure, originally developed by Graham and van der Eb (1973) and later modified by Wigler and co-workers (Wigler *et al.*, 1979), genomic DNA is isolated from donor cells and extracted several times with phenol to remove protein contaminants. This "naked" DNA is then sheared by passage through a 21-gauge needle to obtain DNA fragments of an average size of about 25,000 base pairs (bp). Sheared DNA in a HEPES buffered saline solution is then mixed with an appropriate amount of a $CaCl_2$ solution so that a fine $CaPO_4$ precipitate forms. This DNA/

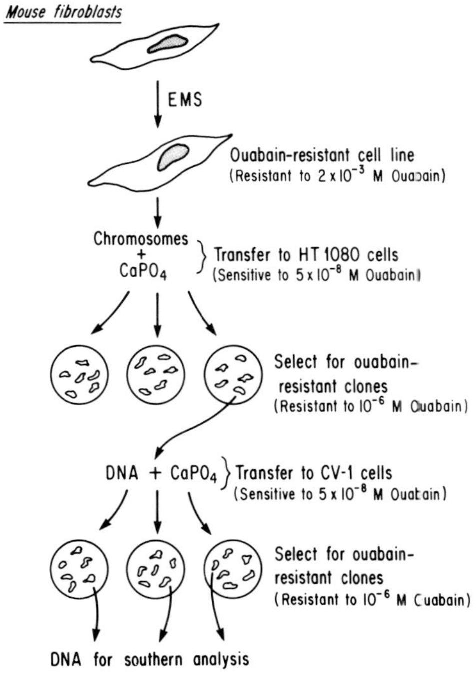

FIG. 1. Transfer of ouabain resistance from mouse to monkey cells. For explanation see text.

$CaPO_4$ precipitate is applied to the recipient cells plated at a density of about 1–2 $\times$ 10^6 cells/10 cm dish. After 4 hr of incubation, cells are "shocked" with a solution of 15% glycerol in HEPES buffered saline. This procedure apparently facilitates uptake of DNA by recipient cells. Cells are then thoroughly washed, grown in culture for 2 days, then subjected to selective pressure with an appropriate drug.

In general, drug-resistant colonies become visible about 2–3 weeks after DNA transfection. These colonies can be harvested and expanded into mass cultures for subsequent analysis. In the experiments described here, DNA from a cell line transformed to ouabain resistance in a primary round of DNA transfection can again be used to transform ouabain-sensitive cells to ouabain resistance in a secondary round of DNA transfection. The purpose of this procedure is to segregate donor DNA sequences associated with ouabain resistance from all other donor DNA sequences, thus reducing the complexity of donor DNA sequences contained within the hybrid ouabain-resistant transformed cell.

The ouabain-resistant transformed cell line HT1080 trans mouse oua^R-F was used as the source of donor DNA in the experiments reported here. This cell line was kindly provided by T. Gross and R. Baker of MIT. HT1080 trans mouse oua^R-F was derived by the transfer of metaphase chromosomes from mouse fibroblasts, which had been selected for resistance to high levels of ouabain following treatment with the mutagen EMS, to ouabain-sensitive human fibrosarcoma HT1080 cells (T. Gross and R. Baker, unpublished results). Genomic DNA from this cell line (resistant to 3×10^{-3} M ouabain) was then prepared in our laboratory and used to transform ouabain-sensitive CV-1 cells by the calcium phosphate precipitation method, as shown in Fig. 1. Several independent DNA

TABLE I

TRANSFER OF OUABAIN RESISTANCE[a]

Donor DNA	Number of colonies/μg DNA	Number of colonies/dish
HT 1080 trans mouse oua^R-F	0.3	12, 12, 9
HT 1080	0	0
CV-1	0	0
HT 1080 trans mouse oua^R-F (*Eco*RI)[b]	0.3	10, 12, 10
pRLoua^R (50 ng)[c]	5/ng	250, 237

[a] About 40 μg of genomic DNAs and 50 ng of plasmid DNA were applied to CV-1 cells. Each dish contained ~2 $\times$ 10^6 cells. For ouabain selection, each dish was trypsinized and replated at ⅓ density (genomic DNA transfection) or ½ density (plasmid DNA transfection) in medium containing 10^{-6} M ouabain. The total number of colonies on the plates represents the transformation efficiency for the amount of added DNA.

[b] DNA was digested to completion with *Eco*RI prior to transfection.

[c] Plasmid DNA was mixed with 40 μg of carrier CV-1 DNA prior to transfection.

transfers of this type consistently yielded secondary ouabain-resistant transformants at a rate of about 0.3/μg of donor DNA, as demonstrated in Table I. When CV-1 or HT1080 DNAs were used as donor DNAs, no ouabain-resistant transformants were ever observed. Ten independent ouabain-resistant secondary transformants were isolated and expanded into mass populations. These cell lines retained the ability to proliferate in ouabain following growth for 10 generations in the absence of ouabain and reexposure to the drug. This result suggests that these transformed cell lines have stably incorporated and expressed the donor ouabain resistance gene.

B. Detection of the Transforming DNA Segment

The method used here to detect and identify the DNA segment responsible for ouabain resistance is based on earlier experiments, which showed that human DNA sequences can be distinguished from mouse DNA in human–mouse hybrid cell lines (Gusella *et al.*, 1980; Shih *et al.*, 1981). This detection is made possible by the fact that the human genome contains about 300,000 or more copies of a sequence that is interspersed widely throughout cellular DNA (Schmid and Deininger, 1975). Thus virtually every human gene is closely linked to one or more copies of this sequence. These repetitive sequence blocks, often referred to as Alu sequences, have diverged sufficiently from their rodent counterparts to allow their specific detection by sequence hybridization to human repeat DNA. This approach has been successfully applied to the identification and isolation of transforming genes transferred by DNA tranfection from human cancer cells to mouse 3T3 fibroblasts (Shih and Weinberg, 1982).

Since Alu-like seqeunces are also present in mouse DNA (Schmid and Jelinek, 1982), it seemed likely to us that one or more of these sequences will be closely linked to the mouse ouabain resistance gene. This sequence would thus serve as a marker in sequence hybridization studies designed to distinguish the transforming mouse DNA segment from CV-1 DNA sequences in ouabain-resistant transformants.

Initial experiments designed to detect the DNA sequences containing the mouse ouabain resistance gene were carried out using a secondary ouabain-resistant cell line, referred to as HTCV-2, which was derived by the procedure outlined in Section II,A. DNA sequence hybridization analysis was carried out by the Southern blot procedure (Southern, 1975). In this method, DNA was cleaved with various restriction enzymes and the DNA fragments resolved by electrophoresis through 1% agarose gel. DNA fragments were transferred to nitrocellulose and sequence hybridization carried out using ^{32}P-total mouse DNA as a probe. The filter was then exposed to X-ray film at −80°C and hybridizing bands visualized by autoradiography. The result of such an experiment is shown

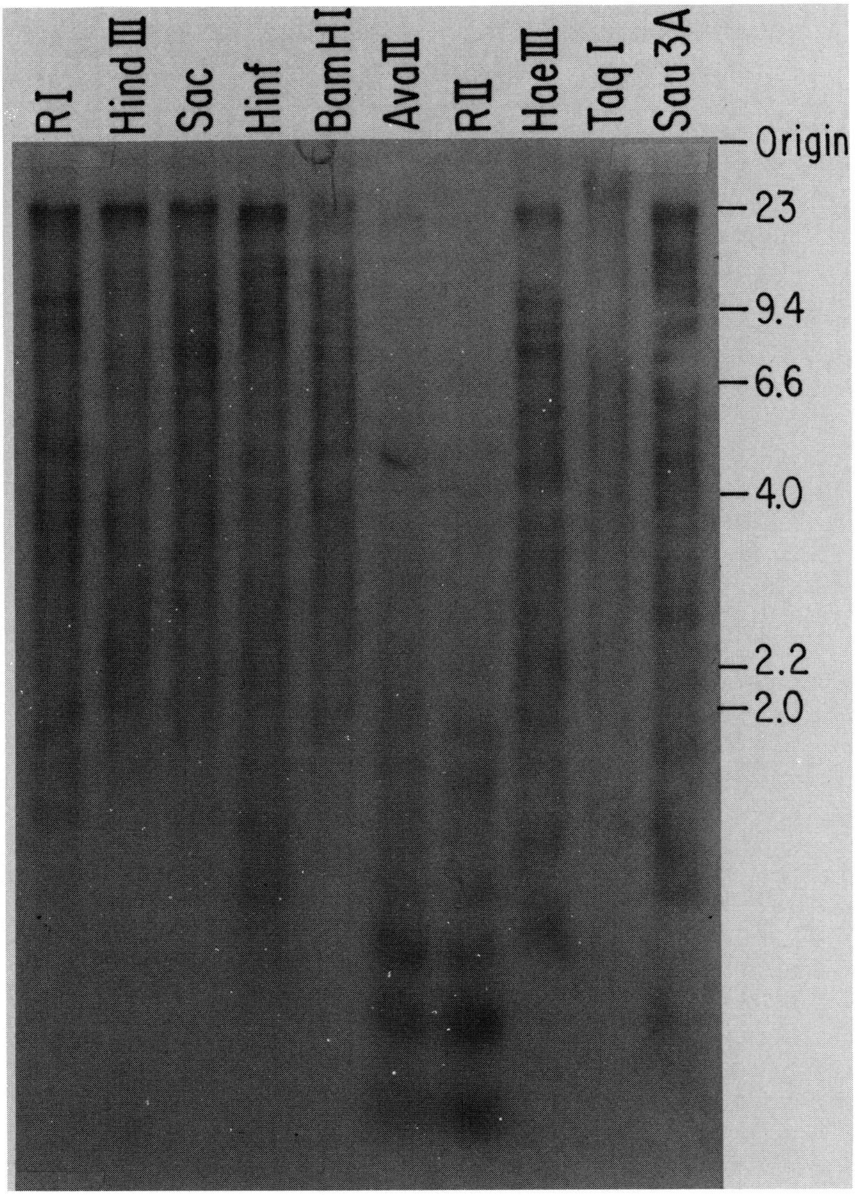

FIG. 2. Southern analysis of a secondary oua^R transformed cell line employing total mouse DNA as probe. DNA from the 2° oua^R transformant HTCV-2 was cleaved with the restriction endonucleases listed in the figure. DNA (10 μg) from each digest was resolved by electrophoresis through a 1% agarose gel. After transfer to nitrocellulose, the blot was reacted with 2×10^7 cpm of ^{32}P-labeled total mouse DNA. Molecular weight markers (in kb) are shown at right.

in Fig. 2. A large number of bands can be seen in most lanes of the filter. The possibility that these bands might represent satellite DNA sequences was established by treatment of DNA with the enzymes *Eco*RII and *Ava*II (Fig. 2, lanes 6 and 7). These enzymes are known to cleave within mouse satellite DNA se-

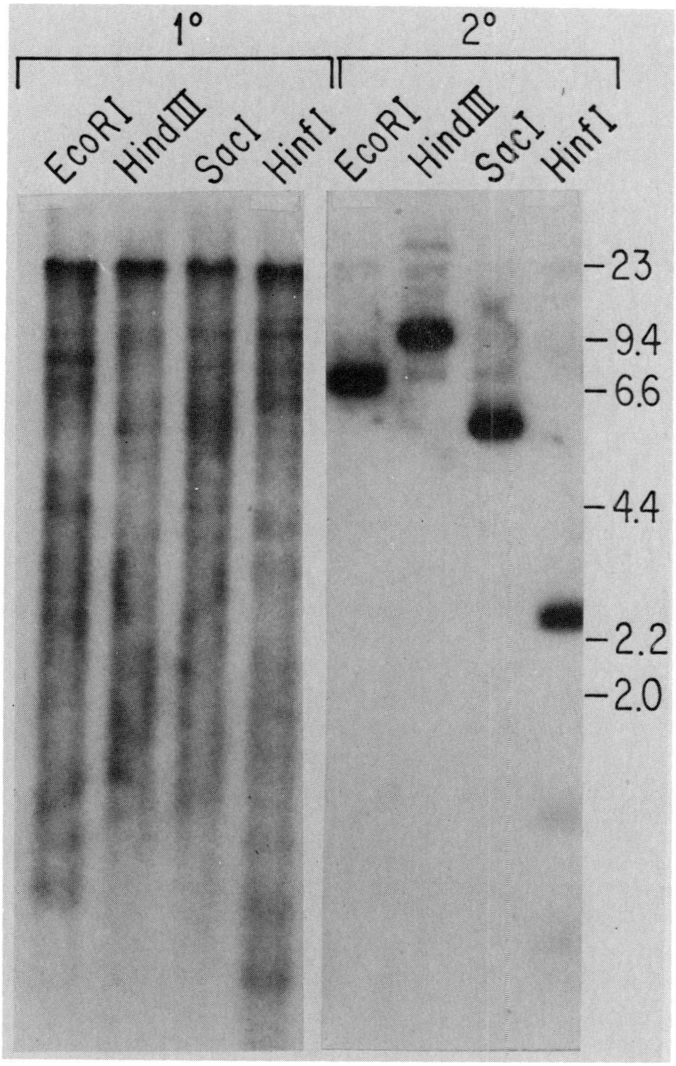

FIG. 3. Southern analysis of primary and secondary oua^R transfectants employing a cloned mouse repeat DNA sequence as probe. DNA from HT1080 trans mouse oua^R-F cells (1° transfectant) and HTCV-2 (2° transfectant) was cleaved with the restriction enzymes described in the figure. DNA (10μg) was run in each lane of a 1% agarose gel. After transfer to nitrocellulose, the blot was reacted with 10^7 cpm of pMR81. Molecular weight markers are shown at the right.

quences (Brown and Dover, 1980). A pattern of bands diagnostic of mouse satellite DNA was observed: a 240-bp sequence and a ladder representing multiples of the basic repeat unit.

This result indicates that preferential hybridization to mouse satellite sequences will obscure the presence of the DNA sequence containing the ouabain resistance gene when total mouse DNA is used as a probe. We therefore screened DNA from ouabain-resistant cell lines with several different mouse repeat DNA sequence probes which Lorraine Albritton had cloned in the plasmid pBR322 (unpublished). It was hoped that one of these probes would react specifically with mouse repetitive sequences closely linked to the ouabain resistance gene. One probe, designated pMR81, appeared to fulfill this criterion.

DNAs from HT1080 trans mouse oua^R-F cells and the ouabain-resistant secondary transformant HTCV-2 were digested with several restriction enzymes and analyzed by sequence hybridization to pMR81 in a Southern blot. As shown in Fig. 3, primary (1°) transfectant HT1080 trans mouse oua^R-F cells containing the mouse ouabain resistance gene also acquired a large number of other mouse DNA sequences. Serial passage of the transforming sequences in a second round of transfection leads to a secondary transfectant (2°), whose DNA carried only a limited number of bands which were reactive with pMR81. CV-1 DNA was not reactive with this probe. The *Eco*RI fragment of about 7 kilobases (kb) represents a characteristic signature of secondary ouabain-resistant transformants. All cells that are transformed to ouabain resistance carry this fragment or one of very similar size (Levenson *et al.*, 1984).

C. Identification of the Transforming DNA Segment

The cosegregation of the *Eco*RI fragment and the repeat DNA sequence homologous to pMR81 indicated that the two elements were closely linked, but did not establish that the functional ouabain resistance DNA segment was contained within the *Eco*RI fragment. An alternative theory would place the ouabain resistance gene on some other DNA fragment that lacks repeat sequences homologous to pMR81 and therefore escapes detection by blot hybridization.

We first wished to determine whether the transforming DNA segment was carried within the *Eco*RI fragment. To do this, DNA from the primary transformant HT1080 trans mouse oua^R-F was subjected to digestion with *Eco*RI prior to transfection into CV-1 cells. When *Eco*RI-cleaved DNA was applied to CV-1 cells, the biological activity of transforming DNA was unaffected, as shown in Table I. This result suggests that transforming activity is contained within an *Eco*RI restriction fragment. When the DNAs of resulting secondary transformants were analyzed for the presence of mouse repeat sequences, all such secondary transformants were found to contain an *Eco*RI fragment of about 7 kb which still contained sequences that reacted positively with pMR81 probe (data

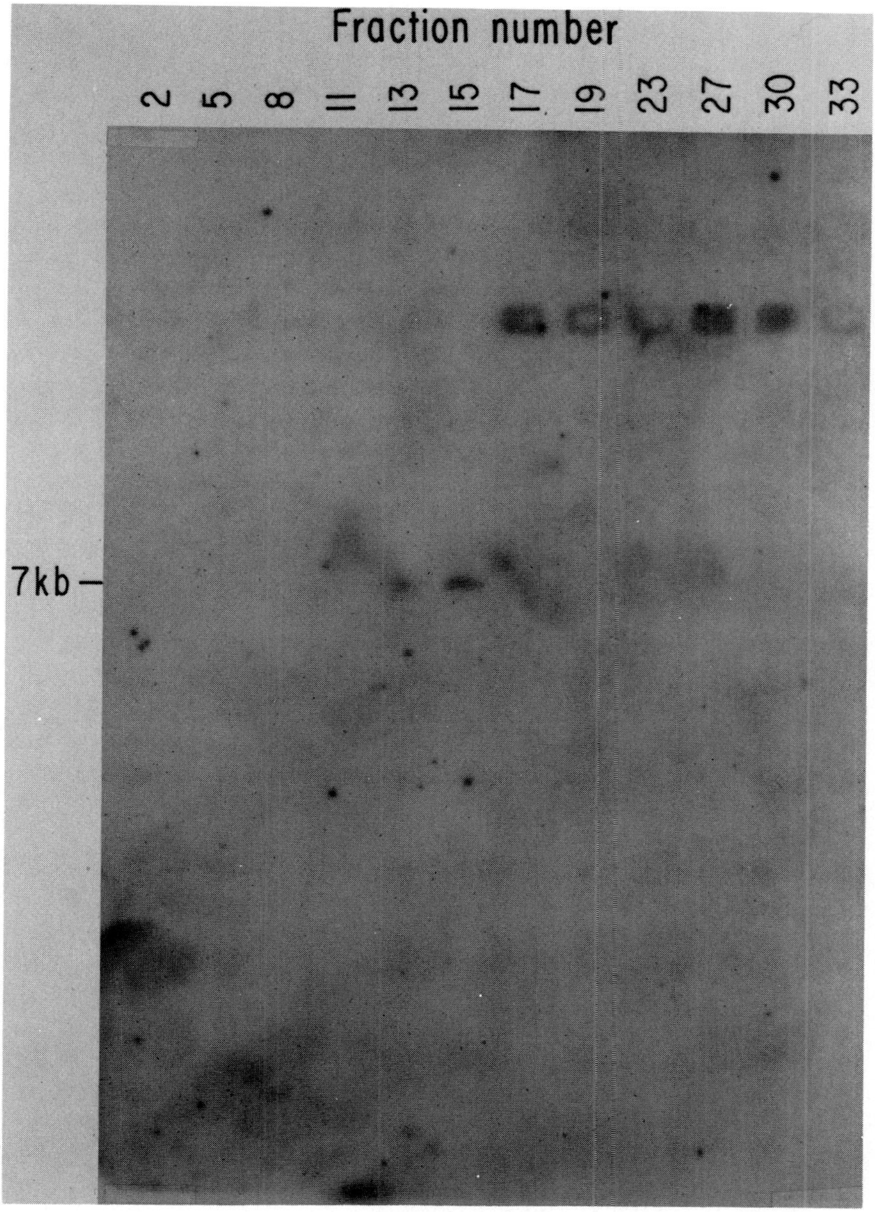

FIG. 4. Southern analysis of sucrose gradient fractionated secondary transfectant DNA. DNA from the oua^R 2° transfectant HTCV-2 was cleaved with *Eco*RI and the DNA fragments separated by centrifugation through a 10–40% sucrose gradient. Aliquots from sucrose gradient fractions were electrophoresed through a 1% agarose gel. After transfer to nitrocellulose, the blot was screened with pMR81 as probe in sequence hybridization analysis.

TABLE II

TRANSFORMING ACTIVITY OF *EcoRI*-CLEAVED SECONDARY TRANSFORMANT DNA FRACTIONS[a]

Fraction	Size (kb)	Hybridization to pMR81	No. colonies/dish
1–7	>15	–	0
8–13	>9	–	0
14–16	5–8	+	10, 11, 8
17–25	2–5	–	0
26–33	<2	–	0
HTCV-2 (unfractionated)			12, 11, 11

[a] DNA from the 2° transformant HTCV-2 was digested to completion with *Eco*RI and fractionated on a 10–40% sucrose gradient. Fractions exhibiting positive and negative hybridization to pMR81 on Southern blots were pooled and used to transfer CV-1 cells. DNAs from the pooled fractions were mixed with the appropriate amount of CV-1 DNA so that 40 μg of DNA was transfected in each case.

not shown). We thus established that cleavage of transforming DNA with *Eco*RI preserves biological activity of the ouabain resistance gene and that treatment with this enzyme does not break the linkage between the *Eco*RI fragment and the repetitive DNA sequence element.

In order to ascertain whether transforming activity and the repetitive sequence element were both contained on the 7 kb *Eco*RI fragment, the following experiment was performed. DNA from the secondary ouabain-resistant transformant HTCV-2 was cleaved with *Eco*RI and the DNA fragments separated by sucrose gradient fractionation. DNA from gradient fractions was resolved by electrophoresis throught a 1% agarose gel. Following transfer to nitrocellulose, sequence hybridization was carried out employing pMR81 as probe. As demonstrated in Fig. 4, only one fraction (No. 14) contained a DNA segment that reacted positively with pMR81. The size of the DNA in this fraction was about 6.5 kb. DNAs from various fractions were then used to transfect CV-1 cells. As shown in Table II, DNA from gradient fractions that failed to hybridize with pMR81 yielded no ouabain-resistant transformants. Only *Eco*RI fragments of 6.5–7 kb that gave positive signals in Southern blots with pMR81 probe were able to transform CV-1 cells to ouabain resistance. This result suggests that the ouabain resistance gene lies within one or more of these *Eco*RI fragments, and that the mouse repeat DNA sequence homologous to pMR81 is closely linked to the ouabain resistance gene.

III. MOLECULAR CLONING OF THE OUABAIN RESISTANCE GENE

A. Construction and Screening of a λ Phage Library

The strategy for molecular cloning of the ouabain-resistance gene began with the insertion of *Eco*RI fragments of the secondary ouabain-resistant transformant

HTCV-2 into an appropriate λ phage vector. We then planned to screen the resultant recombinant phage library for those components that exhibited sequences reactive with pMR81 probe. Sucrose gradient fractions of *Eco*RI-cleaved HTCV-2 DNA that exhibited transforming activity (described in Section II,C) following transfection into CV-1 cells were pooled together. This enriched fragment mixture was ligated to *Eco*RI-cleaved λCharon 16A bacteriophage DNA and packaged *in vitro* (Grosveld *et al.*, 1981). Approximately 10^6 plaque-forming units were mixed with *E. coli* strain LE392 and plated. The resultant phage plaques were screened by hybridization to pMR81 according to the procedure of Benton and Davis (1977). Twelve phage stocks were isolated that retained reactivity with pMR81 through four cycles of plaque purification and rescreening. These 12 phage stocks were then plated in groups of twos and threes (29 total plaques) in a grid assay and rescreened a fifth time with pMR81. As shown in Fig. 5, all 12 phage stocks retained reactivity with pMR81. Plaques formed by λCharon 16A that did not contain inserted DNA, or contained inserted CV-1 DNA, failed to react with the mouse probe (faint areas seen on the filter).

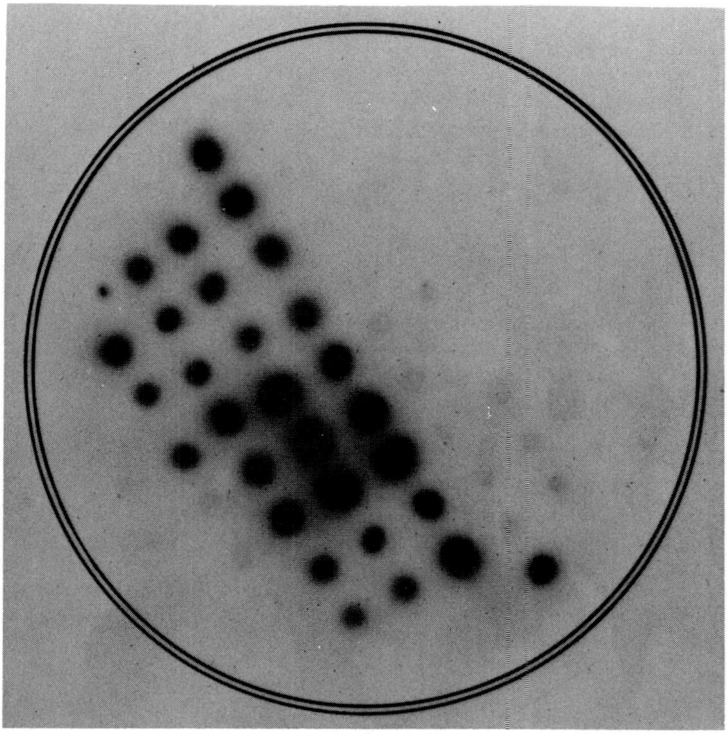

FIG. 5. Plaque hybridization of recombinant phage. Phages containing the 6.5-kb mouse insert were plated in groups of twos and threes. Phages containing no inserted DNA or inserted CV-1 DNA were plated as controls. Phage plaques were transferred to nitrocellulose and the filter reacted with 2 × 10^6 cpm of ^{32}P-labeled pMR81.

DNA was prepared from 4 of the 12 phage stock and all were found to contain identical inserted DNAs, all 6.5 kb in size (data not shown). Each of these four DNAs was applied to CV-1 cells and selection with 10^{-6} M ouabain carried out. After 4 days of selection, about 50 colonies were observed in each of the four dishes, whereas no colonies were observed in a control dish in which λCharon 16A DNA alone was used as donor DNA. These colonies proved to be highly unstable, however. Only one ouabain-resistant transformant survived under selective pressure. When the DNA of this transformant was analyzed by sequence hybridization to pMR81 in a Southern blot, it did not appear to contain the 6.5-kb mouse DNA fragment. It is thus possible that either the 6.5-kb mouse DNA insert did not contain transforming activity or that the transfected DNA was not stably integrated into the genome of the recipient cells.

B. Subcloning of Inserted DNA into a Plasmid Vector: Isolation of a Biologically Active Subclone

Recent data have shown that phage DNA may act to inhibit mammalian cell transformation when mixed with eukaryotic DNA (Yoder and Genesan, 1983). We therefore subcloned the *Eco*RI 6.5-kb DNA insert purified from phage into the plasmid vector pAT153. (pAT153 is a derivative of pBR322: its utility as a cloning vector derives from the fact that a 700-bp sequence which inhibits expression of transforming DNA has been deleted.) To do this, DNA from one of the four phage stocks containing 6.5-kb inserted DNA was digested with *Eco*RI and the DNA fragments separated by electrophoresis through a 0.6% agarose gel. The 6.5-kb fragment was isolated from the gel and ligated to *Eco*RI-cleaved pAT153. This DNA was used to transform *E. coli* strain HB101. DNA from this subclone, pRLoua^{R}, was used to transfect CV-1 cells. As shown in Table I, this subclone has a high degree of biological activity upon transfection. Plasmid DNA (50 ng) yielded about 250 ouabain-resistant colonies. Sequence hybridization of DNAs from 5 of these colonies to pMR81 indicated that these DNAs contained both plasmid and inserted DNA sequences (data not shown). This result provides direct evidence that the ouabain resistance (*oua*^{R}) gene is contained within the 6.5-kb *Eco*RI segment present in pRLoua^{R}.

C. Physical and Functional Map of the Transforming DNA Segment

A restriction map of the DNA fragment containing the ouabain resistance gene was constructed by a series of single and double restriction endonuclease digestions. The map of the transforming DNA segment is shown in Fig. 6. One issue we wished to address concerned the location of the repeat sequence block homol-

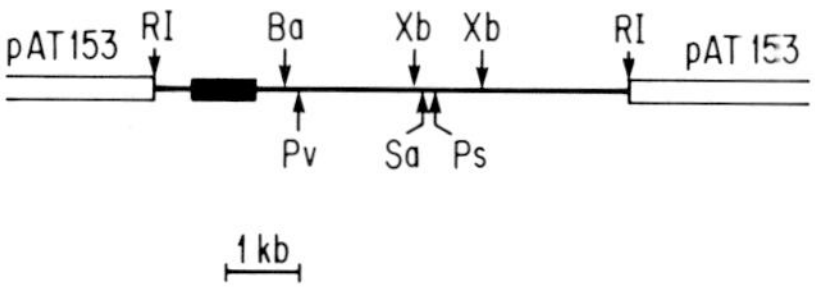

FIG. 6. Restriction map of the *oua*^{R} gene. pRLoua^R was cleaved with *Eco*RI and the fragments resolved by electrophoresis through a 0.6% agarose gel. The 6.5-kb insert was purified from the gel and mapped by a series of single and double digestions. Ba, *Bam*HI; Sa, *Sac*I; Pv, *Pvu*II; Xb, *X bal*; Ps, *Pst*I. The shaded box designates the location of repeat sequences homologous to pMR81.

ogous to pMR81. To do this, the inserted 6.5-kb DNA fragment of pRLoua^R was treated with a series of single and double restriction endonucleases. The resulting restriction fragments were resolved by electrophoresis through a 1.5% agarose gel, transferred to nitrocellulose, and tested for their reactivity with pMR81 in a Southern blot. When *Bam*HI was used to cleave the 6.5-kb DNA sequence, two restriction fragments of 1.5 kb and 5 kb were obtained. Only the 1.5-kb sequence contained sequences reactive with pMR81. In fact, only those restriction fragments that included the 1.5-kb *Eco*RI–*Bam*HI segment exhibited hybridization to the pMR81 probe. These results support the view that the repeat sequence block homologous to pMR81 must be located within the 1.5-kb *Eco*RI–*Bam*HI segment of the cloned insert.

The construction of a physical map of the transforming DNA segment has allowed us to address a further issue of significance, the derivation of a functional map of the transforming DNA segment. Put another way, we wished to determine whether a portion of the 6.5-kb DNA insert of pRLoua^R contained all the information required to transform cells to ouabain resistance, or if the entire 6.5-kb DNA sequence was essential for biological activity. To approach this problem we examined the ability of various segments of the 6.5-kb DNA insert to transform CV-1 cells to ouabain resistance. Purified 6.5-kb DNA insert was digested with *Bam*HI, *Sac*I, or *Bam*HI/*Sac*I. The DNA from each reaction mixture was applied to CV-1 cells and cells were maintained under selective pressure with 10^{-6} *M* ouabain. As shown in Table III, purified 6.5-kb DNA was active in transforming CV-1 cells to ouabain resistance. *Sac*I and *Bam*HI/*Sac*I digestion destroyed the ability of 6.5-kb DNA to transform cells to ouabain resistance. Treatment with *Bam*HI, on the other hand, preserved the biological activity of the transforming DNA segment. These data suggest that the ouabain resistance gene is contained within the 5-kb *Eco*RI-*Bam*HI segment of the inserted DNA. Although visual inspection of the *Bam*HI restriction digest indicated that digestion was complete, it is possible that a small number of intact 6.5-kb DNA molecules may have been present in the reaction mixture used to transfect CV-1 cells, thus accounting for the ability of this reaction mixture to yield ouabain-resistant transformants. To rule out this possibility, we purified the 1.5- and 5-kb

TABLE III

TRANSFORMING ACTIVITY OF THE *oua*^{R} GENE AND *oua*^{R} GENE SEGMENTS[a]

Donor DNA	Enzyme	No. colonies/dish
pRLoua^R	–	203, 218
6.5-kb insert[b]	–	153, 162
6.5-kb insert	*Sac*I	0
6.5-kb insert	*Bam*HI/*Sac*I	0
6.5-kb insert	*Bam*HI	136, 123
EcoRI–*Bam*HI (1.5 kb)[c]		0
*Eco*RI–*Bam*HI (5 kb)[c]		77, 63

[a] In each case, 50 ng of donor DNA mixed with 10 μg of CV-1 DNA and the mixture cotransfected into CV-1 cells. Selection was carried out at 10^{-6} *M* ouabain. Each dish was replated at ½ density prior to transfection.

[b] pRLoua^R was digested with *Eco*RI and the 6.5-kb insert purified from an agarose gel.

[c] 6.5-kb insert was digested with *Bam*HI and the 1.5- and 5-kb segments purified from an agarose gel.

*Bam*HI restriction fragments from an agarose gel and separately transfected these purified DNA fragments into CV-1 cells. The data presented in Table III indicate that the ability to transform CV-1 cells to ouabain resistance resides solely within the 5-kb *Eco*RI–*Bam*HI segment of inserted DNA. We have not as yet determined whether a portion of the biologically active 5-kb segment is functionally irrelevant.

IV. ANALYSIS OF OUABAIN RESISTANCE GENE TRANSCRIPTS

Preliminary experiments designed to characterize ouabain resistance gene transcripts in mouse tissue have been carried out employing the cloned mouse ouabain resistance (*oua*^{R}) gene as a hybridization probe. Our initial interest here is focused on the size of the mRNA sequence transcribed from this gene. This approach is of significance for determining whether the *oua*^{R} gene could be the DNA sequence coding for the α subunit of the Na^+,K^+-ATPase. The α subunit has, in most systems, an apparent MW of about 100,000 (Cantley, 1981). A protein of this size would thus contain about 1000 amino acids, although the exact number is unknown. Therefore, approximately 3000 bases of RNA would be required to encode the complete amino acid sequence of the α subunit. The ability of the cloned *oua*^{R} gene to hybridize with an mRNA of about 3000 bases would support the initial assumption that the DNA sequence responsible for conferring ouabain resistance in transfection experiments is the sequence coding

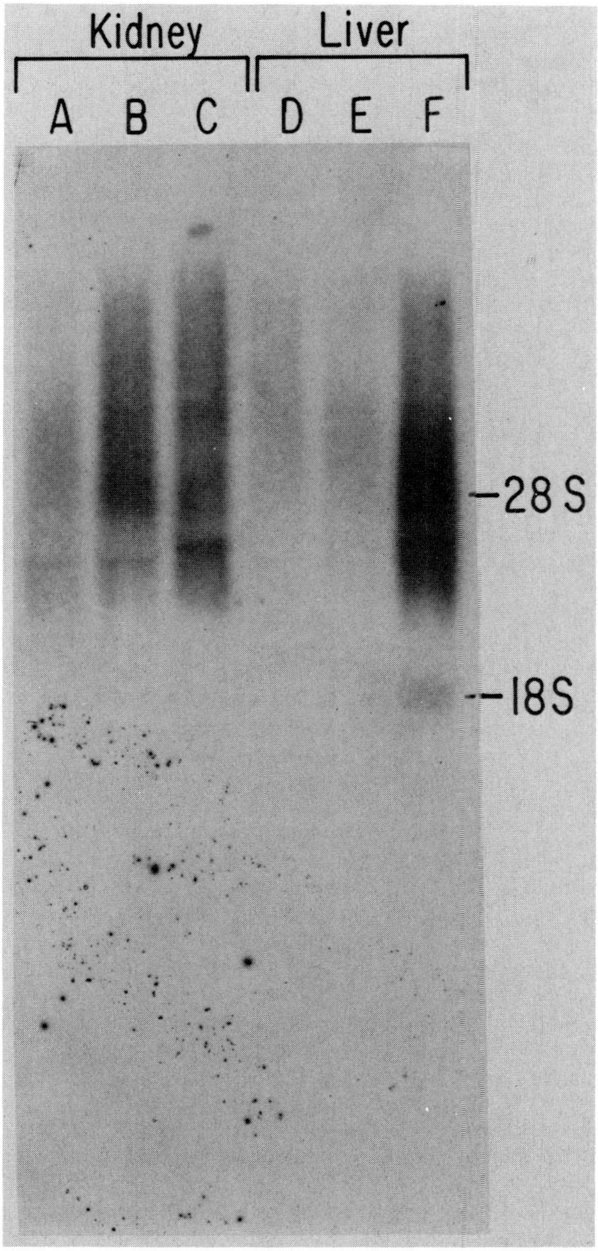

FIG. 7. Northern analysis of *oua*^{R} gene transcripts in mouse tissues. RNAs from mouse kidney and liver were fractionated on a formaldehyde agarose gel. (A and D) 30 μg poly(A)$^-$ RNAs; (B and E) 30 μg total cell RNAs. (C and F) 10 μg poly(A)$^+$ RNAs. After transfer to nitrocellulose, the blot was reacted with 1.5 × 10^7 cpm ^{32}P-labeled pRLoua^R DNA. Molecular weight markers are at right.

for the α subunit of the Na^+,K^+-ATPase. From a general perspective, the approach described in this section is also of importance to the molecular analysis of the ouabain resistance locus. For example, it will be possible to measure the levels of *oua*^{R} gene transcripts in tissues such as brain and kidney which appear to have greater than 100-fold more Na^+,K^+-ATPase activity than other tissues (Sweadner and Goldin, 1980). Of further significance, the analysis of *oua*^{R} gene transcripts in secondary and tertiary ouabain-resistant transformed cell lines should provide insight into the molecular mechanisms involved in the achievement of ouabain resistance and regulation of *oua*^{R} gene expression.

RNAs from mouse liver and kidney were analyzed by sequence hybridization in Northern blots. Total cellular RNA was isolated from minced tissues by the guanidine isothiocyanate method (Chirgwin *et al.*, 1979) and selected on oligo(dT)-cellulose (Aviv and Leder, 1972). RNA was fractionated on a denaturing formaldehyde–agarose gel (Goldberg, 1980) and transferred to nitrocellulose according to the method of Thomas (1980). Sequence hybridization was carried out with ^{32}P-pRLoua^R as probe. As shown in Fig. 7, both kidney and liver poly(A)$^+$ RNAs (lanes C and F) exhibit an intense smear of hybridization from about 22 S to greater than 35 S in size. Kidney poly(A)$^-$ RNA (lane A) and total cell RNA (lane B) show less intense but similar patterns of hybridization. Liver poly(A)$^-$ RNA (lane D) and total cell RNA (lane E) exhibit relatively faint smears of hybridization.

The high background of hybridization in this experiment prevents the visualization of unique mRNA species of potential interest. The most likely explanation for this smearing pattern is that repeat sequences in the probe (pRLoua^R contains the repeat DNA sequence block) hybridize to repeat sequences present within the population of cellular mRNA molecules. Similar results have been described in several other systems in which probes containing repeat sequences have been used to analyze cellular RNA (Soriano *et al.*, 1983; Zucker and Lodish, 1981; Sharp, 1983; Lerman *et al.*, 1983). A 5-kb unique segment of pRLoua^R has been subcloned (Section III,C) and will be employed to probe Northern transfers in order to determine the size of ouabain-resistant gene transcripts and to study the control of expression of the *oua*^{R} gene.

V. AMPLIFICATION OF OUABAIN RESISTANCE DNA SEQUENCES IN TRANSFORMED CELL LINES

Southern analysis of the transforming DNA segment in secondary ouabain-resistant transformed cell lines suggested a possible amplification of the transfected mouse ouabain resistance gene. The transforming mouse DNA bands of 6.5 kb exhibited an intensity of hybridization greater than that expected for a single copy DNA sequence. The experiments described in this section are consistent with the idea that the inserted *oua*^{R} gene is amplified in drug-treated cells and suggest a molecular mechanism for achievement of ouabain resistance by cells in culture.

DNA was prepared from the following cell lines: wild-type CV-1 cells; secondary ouabain-resistant transformed cell lines HTCV-2 and HTCV-4; CV λ 16A-20, a ouabain-resistant cell line that had been treated with λ Ch16A containing inserted 6.5-kb mouse DNA (Section III,A). These DNAs were cleaved with *Eco*RI and 10 μg of each DNA was electrophoresed through a 1% agarose gel. After transfer to nitrocellulose, the filter was probed with ^{32}P-labeled 6.5 kb ouabain resistance DNA isolated from pRLoua^R. The results of this analysis are presented in Fig. 8. Two faintly hybridizing bands of 10 and 4 kb are visible in CV-1 DNA (lane D). These bands represent monkey sequences homologous to the mouse ouabain resistance gene. The 6.5-kb mouse ouabain resistance band exhibits very intense hybridization in HTCV-2 DNA (lane B) and HTCV-4 DNA (lane C). The intensity of hybridization suggests that the mouse ouabain resistance gene has been amplified to high copy number in both HTCV-2 and HTCV-4 cells. The 10- and 4-kb CV-1 DNA sequences are also visible in these secondary ouabain-resistant cell lines. Surprisingly, these bands are present in greater copy number than in wild-type CV-1 cells. As described previously in Section III,A, CVλCh16A-20 DNA (lane A) does not appear to contain the 6.5-kb mouse ouabain resistance gene. However, the 10- and 4-kb bands diagnostic of CV-1 cells are visible and appear to be amplified to a significant extent.

The results presented here are significant for a number of reasons. The ability of the mouse *oua*^{R} gene to detect CV-1 DNA sequences that are amplified in ouabain-resistant CV-1 cell lines strongly supports the initial conclusion that the *oua*^{R} gene has been isolated. Furthermore, the failure to detect mouse oua^R DNA sequences in CVλCh16A-20 cells by Southern blotting suggests that this cell line has achieved drug resistance by virtue of amplification of endogenous *oua*^{R} gene sequences. Thus in DNA transfection experiments, ouabain resistance may be achieved by integration and amplification of transfected mouse oua^R DNA sequences, and/or amplification of host CV-1 *oua*^{R} gene sequences.

The data presented here indicate that gene amplification may be an important mechanism for the achievement of ouabain resistance in cell culture. As reported elsewhere (Schimke, 1982), the phenomenon of drug resistance as conferred by gene amplification may be a common response of cells to cytotoxic selection with drugs. For example, the resistance of cells to methotrexate or cadmium has been shown to result from the amplification of the genes for dihydrofolate reductase (Alt *et al.*, 1978) and metallothionein-I (Beach and Palmiter, 1981), respectively. In both cases, gene amplification leads to the overproduction of intracellular proteins that are the specific targets of the inhibitors used.

The ability of CV-1 cells to achieve ouabain resistance following transfection of mouse DNA sequences may result solely from the amplification and ex-

FIG. 8. Amplification of oua^R DNA sequences in transformed cell lines. Genomic DNAs were cleaved with *Eco*RI, resolved by electrophoresis through a 1% agarose gel, and transferred to nitrocellulose. DNA (10 μg) was run in each lane. DNAs were prepared from the following cell lines: CVλ16A-20 (A), derived from transfection of phage DNA containing the 6.5-kb insert; HTCV-2 (B), a 2° transfectant; HTCV-4 (C), a 2° transfectant; CV-1 (D). The blot was reacted with 10^7 cpm of pRLoua^R DNA. The arrow indicates the mouse *Eco*RI 6.5-kb DNA fragment.

pression of the mouse *oua*^{R} gene. On the other hand, amplification of endogeneous CV-1 sequences in transformed cell lines suggests that these sequences may play an important role in the establishment of ouabain resistance as well. It is possible that transfected mouse *oua*^{R} gene sequences may serve to transiently "rescue" CV-1 cells from ouabain cytotoxicity. Amplification of the mouse *oua*^{R} gene may serve to induce amplification of CV-1 *oua*^{R} gene sequences which in turn leads to the establishment of ouabain resistance. In either case, it will be extremely interesting to determine if gene amplification leads to an overproduction of *oua*^{R} gene transcripts and an increase in the level of Na^+,K^+-ATPase α chains in ouabain-resistant transformed cell lines.

VI. CONCLUSIONS AND PROSPECTS

The isolation of the ouabain resistance gene should provide an important focus for the genetic analysis of the Na^+,K^+-ATPase. The primary goal of future studies will be to establish that the ouabain resistance gene is the sequence coding for the α subunit of the Na^+,K^+-ATPase. Definitive proof of this relationship will depend on either hybridization selection of mRNA followed by translation in a cell-free system or direct comparison of DNA sequence to amino acid sequence.

Isolation of the α subunit gene will permit several important goals to be realized with respect to Na^+,K^+-ATPase structure. It should be possible to determine the complete nucleotide sequences of the cloned α subunit gene and to translate this information into a complete amino acid sequence of the α-subunit protein. The determination of the primary structure of the α subunit should provide a more precise picture of the way the Na^+,K^+-ATPase operates. A further issue of significance is whether multiple forms of the Na^+,K^+-ATPase that occur in different tissues or under different physiological conditions within the same cell exhibit primary sequence differences in the structure of the α subunit. It has been demonstrated that two forms of the Na^+,K^+-ATPase are present in brain tissue (Sweadner, 1979) and in ouabain-resistant transformed cell lines (English, Cantley, and Levenson, unpublished observations). The structural basis for these differences has not been determined, however. The isolation of the α subunit gene opens the possibility of using nucleotide mapping and sequencing techniques to address this question. Comparison of the amino acid sequence of the mouse α-chain sequence to the primate α-chain amino acid sequence could give insight into the differences in the ouabain binding site between the mouse and primate enzymes. This approach combined with point mutation techniques (Shortle *et al.*, 1981) should lead to the precise identification of the ouabain binding site within the Na^+,K^+-ATPase.

The ability to insert the α-chain gene in a mammalian expression vector will eventually facilitate experiments designed to examine the effects of other genetic loci on the α-chain gene by making it technically simpler to transfer the α-chain gene into and out of mammalian cells. In the long term, the ability to introduce the α-chain gene under the control of an inducible promoter into a mammalian cell opens the possibility of regulating the level of α subunits within the cell in a controlled manner. Regulation of α-subunit gene transcription and α-subunit protein levels would make it possible to observe the physiological consequences of increasing Na^+,K^+-ATPase levels on other aspects of cell behavior and metabolism.

The approach described here for the isolation of the ouabain resistance gene has the potential to be applied to other polypeptide chains regulating ion levels in the cell. The use of DNA transfer in combination with recombinant DNA techniques could provide a powerful approach for the isolation of the DNA sequences coding for other ion transport proteins, especially in situations where specific drug-resistant mutants can be selected. In cases where gene amplification leads to drug resistance, newly developed techniques (Roninson, 1983) may provide a direct route for the isolation of amplified DNA sequences which code for ion transport proteins.

ACKNOWLEDGMENTS

I am indebted to David Housman (MIT) for his invaluable contributions to this work. I thank T. Gross and R. Baker for providing the primary oua^R cell lines used in these studies, and V. Racaniello, L. Albritton, and J. Tamkun for their contributions to various parts of this work. I am grateful to N. Crawford, K. Kabnick, H. Bernstein, and P. Sarnow for their helpful comments on the manuscript, and to K. Thurmond and M. B. Hicks for expert secretarial assistance. This project was supported by a grant from the NCI (to D. Housman) and a New Investigator Grant from the Cystic Fibrosis Foundation (to R. Levenson).

REFERENCES

Alt, F. W., Kellems, R. M., Bertino, J. R., and Schimke, R. T. (1978). Selective multiplication of dihydrofolate reductase genes in methotrexate-resistant variants of cultured murine cells. *J. Biol. Chem.* **253,** 1357–1370.

Aviv, H., and Leder, P. (1972). Purification of biologically active globin messenger RNA by chromotography on oligothymidylic acid-cellulose. *Proc. Natl. Acad. Sci. U.S.A.* **69,** 1408–1412.

Baker, R. M. (1976). Genetic and cellular properties of ouabain resistant mutants. *In* "Biogenesis and Turnover of Membrane Macromolecules" (J. S. Cook, ed.), pp. 93–103. Raven, New York.

Beach, L. R., and Palmiter, R. D. (1981). Amplification of the metollothionein-I gene in cadmium-resistant mouse cells. *Proc. Natl. Acad. Sci. U.S.A.* **78,** 2110–2114.

Benton, W. D., and Davis, R. W. (1977). Screening λgt recombinant clones by hybridization to single plaques *in situ*. *Science* **196,** 180–182.

Brown, S. D. M., and Dover, G. A. (1980). The specific organization of satellite DNA sequences on the X-chromosome of Mus. musculus: Partial independence of chromosome evolution. *Nucleic Acids Res.* **8,** 781–792.

Cantley, L. C. (1981). Structure and mechanism of the (Na,K)-ATPase. *Curr. Top. Bioenerg.* **11,** 201–237.

Chirgwin, J. M., Przybyla, A. E., MacDonald, R. J., and Rutter, W. J. (1979). Isolation of biologically active ribonucleic acid from sources enriched in ribonuclease. *Biochemistry* **18,** 5294–5299.

Corsaro, C. M., and Pearson, M. L. (1981). Competence for DNA transfer of ouabain resistance and thymidine kinase: Clonal variation in mouse L-cell recipients. *Somat. Cell Genet.* **7,** 617–630.

Goldberg, D. A. (1980). Isolation and partial characterization of the Drosophila alcohol dehydrogenase gene. *Proc. Natl. Acad. Sci. U.S.A.* **77,** 5794–5798.

Graham, F. L., and van der Eb, A. J. (1973). A new technique for the assay of infectivity of human adenovirus-5 DNA. *Virology* **52,** 456–467.

Grosveld, F. G., Dahl, H.-H. M., de Boer, E., and Flavell, R. A. (1981). Isolation of β-globin related genes from a human cosmid laboratory. *Gene* **13,** 227–237.

Gusella, J. F., Keys, C., Varsanyi-Breiner, A., Kao, F., Jones, C., Puck, T. T., and Housman, D. (1980). Isolation and localization of DNA segments from specific human chromosomes. *Proc. Natl. Acad. Sci. U.S.A.* **77,** 2829–2833.

Kozak, C. A., Fournier, R. E. K., Leinwand, L. A., and Ruddle, F. H. (1979). Assignment of the gene governing cellular ouabain resistance to Mus. musculus chromosome-3 using human/mouse microcell hybrids. *Biochem. Genet.* **17,** 23–34.

Levenson, R., Racaniello, V. R., Albritton, L., and Housman, D. (1984). Molecular cloning of the mouse ouabain resistance gene. *Proc. Natl. Acad. Sci. U.S.A.* **81,** 1489–1493.

Lerman, M. I., Thayer, R. E., and Singer, M. F. (1983). Kpn-I family of long intersperred repeated DNA sequences in primates: Polymorphism of family members and evidence for transcription. *Proc. Natl. Acad. Sci. U.S.A.* **80,** 3966–3970.

Robbins, A. R., and Baker, R. M. (1977). (Na, K)-ATPase activity in membrane preparations of ouabain-resistant HeLa cells. *Biochemistry* **16,** 5163–5168.

Roninson, I. B. (1983). Detection and mapping of homologous, repeated and amplified DNA sequences by DNA renaturation in agarose gels. *Nucleic Acids Res.* **11,** 5413–5431.

Ruoho, A., and Kyte, J. (1974). Photoaffinity labeling of the ouabain-binding site on (Na^{+} + K^{+}) adenosine triphosphatase. *Proc. Natl. Acad. Sci. U.S.A.* **71,** 2352–2356.

Schimke, R. T. (1982). Summary. *In* "Gene Amplification" (R. T. Schimke, ed.), pp. 317–334. Cold Spring Harbor Laboratory, Cold Spring Harbor, New York.

Schmid, C. W., and Deininger, P. L. (1975). Sequence organization of the human genome. *Cell* **6,** 345–358.

Schmid, C. W., and Jelinek, W. R. (1982). The Alu family of dispersed repetitive sequences. *Science* **216,** 1065–1070.

Sharp, P. A. (1983). Conversion of RNA to DNA in mammals: Alu-like elements and pseudogenes. *Nature (London)* **301,** 471–472.

Shih, C., and Weinberg, R. A. (1982). Isolation of a transforming sequence from a human bladder carcinoma line. *Cell* **29,** 161–169.

Shih, C., Padhy, L. C., Murray, M., and Weinberg, R. A. (1981). Transforming genes of carcinomas and neuroblastomas introduced into mouse fibroblasts. *Nature (London)* **290,** 261–264.

Shortle, D., DiMaio, D., and Nathans, D. (1981). Directed mutagenesis. *Annu. Rev. Genet.* **15,** 265–294.

Soriano, P., Mennier-Rotival, M., and Bernardi, G. (1983). The distribution of interspersed repeats is non-uniform and conserved in the mouse and human genomes. *Proc. Natl. Acad. Sci. U.S.A.* **80,** 1816–1820.

Southern, E. M. (1975). Detection of specific sequences among DNA fragments separated by gel electrophoresis. *J. Mol. Biol.* **98,** 503–518.

Sweadner, K. J. (1979). Two molecular forms of (Na^+/K^+)-stimulated ATPase in brain: Separation and difference in affinity for strophanthidin. *J. Biol. Chem.* **254,** 6060–6067.

Sweadner, K. J., and Goldin, S. M. (1980). Active transport of sodium and potassium ions. *New Engl. J. Med.* **302,** 777–783.

Thomas, P. S. (1980). Hybridization of denatured RNA and small DNA fragments transferred to nitrocellulose. *Proc. Natl. Acad. Sci. U.S.A.* **77,** 5201–5205.

Wigler, M., Pellicer, A., Silverstein, S., Axel, R., Urlaub, G., and Chasin, L. (1979). DNA-mediated transfer of the adenine phosphoribosyltransferase locus into mammalian cells. *Proc. Natl. Acad. Sci. U.S.A.* **76,** 1373–1376.

Yoder, J. I., and Ganesan, A. T. (1983). Procaryotic genomic DNA inhibits mammalian cell transformation. *Mol. Cell Biol.* **3,** 956–959.

Zucker, C., and Lodish, H. F. (1981). Repetitive DNA sequences cotranscribed with developmentally regulated Dictyostelium discoideum mRNAs. *Proc. Natl. Acad. Sci. U.S.A.* **78,** 5386–5390.

Index

A

B

C

D

E

H

I

R

S

T

Contents of Recent Volumes

Volume 16

Electrogenic Ion Pumps

Volume 19

Structure, Mechanism, and Function of the Na/K Pump

Volume 20

Molecular Approaches to Epithelial Transport

Volume 21

Ion Channels: Molecular and Physiological Aspects

Volume 22

The Squid Axon